Throughout history, humans have sought to make sense of seemingly random and frequently deadly natural disasters. Scientists struggling to develop theories on earthquakes, hurricanes, and epidemics must today confront the possibility that there may be fundamental limits on human comprehension and prediction of nature's intermittent tantrums. *Perils of a Restless Planet* examines our attempts to understand and anticipate these devastating natural phenomena, focusing on the interactions among basic scientific inquiry, technological innovation, and ultimately public policy.

Drawing upon studies of actual events from ancient to present times, the author provides perspectives on a selection of natural disasters and the scientific process of inquiry that has led to progress in understanding them. He draws attention to the scientific challenges that remain, the socioeconomic factors that influence what scientific questions may be studied in the future, and the prospects for achieving a level of scientific understanding that someday may permit us to predict, and ideally mitigate, natural disasters. Extrapolating from the history of disasters, the author suggests what new perils may lie ahead, and how we may someday learn to protect the vulnerable populations on this restless Earth.

This book is written in an informative and accessible way, acknowledging without sensationalizing the gruesome nature of disasters that is a natural fascination for humans. Looking as much at what we don't know as at what we do know, the author takes the reader on a journey where the human quest for understanding confronts the most fearsome perils of our restless planet.

Perils of a Restless Planet

Perils of a
Restless Planet

Scientific Perspectives
on Natural Disasters

ERNEST ZEBROWSKI, JR.

Pennsylvania College of Technology
The Pennsylvania State University

CAMBRIDGE
UNIVERSITY PRESS

PUBLISHED BY THE PRESS SYNDICATE OF THE UNIVERSITY OF CAMBRIDGE
The Pitt Building, Trumpington Street, Cambridge, United Kingdom

CAMBRIDGE UNIVERSITY PRESS
The Edinburgh Building, Cambridge CB2 2RU, UK   http://www.cup.cam.ac.uk
40 West 20th Street, New York, NY 10011-4211, USA   http://www.cup.org
10 Stamford Road, Oakleigh, Melbourne 3166, Australia
Ruiz de Alarcón 13, 28014 Madrid, Spain

First published 1997
Reprinted 1998
First paperback edition 1999

Printed in the United States of America

Typeset in Ehrhardt

A catalog record for this book is available from the British Library

Library of Congress Cataloging in Publication data is available

ISBN 0 521 57374 2 hardback
ISBN 0 521 65488 2 paperback

My thanks to Tina Tickle who prepared the line art.

To my son, David,
my daughter, Angel,
and all the men and women of their generation

Contents

Preface

Our psychological profiles as humans draw us to natural disasters, some as relief workers, others as public servants or engineers whose vision is to prevent future catastrophes, and many as intellectually curious onlookers. Millions of us watch news telecasts of the latest natural disaster with an uneasy combination of awe and the personal sense of relief that we ourselves are not being counted among the victims. In the aftermath, scientists sift through the available data and ask themselves two fundamental questions: (1) What led Mother Nature to behave this way? and (2) How might we humans better anticipate such an event the next time around? These questions, our current and past attempts to answer them, and the scientific challenges that still confront us are continuous threads running through this book.

Clearly, the scientific study of disasters is much more than a purely intellectual exercise. As scientific models of earthquakes, floods, hurricanes, and epidemics are created and validated, they quickly work their way into engineering, medicine, and other professions. Improved scientific understanding also filters into the public policy arena, resulting, for instance, in amended construction codes and improved evacuation planning. A long list of beneficial spin-offs of disaster science could be put together here; this said, I'll spare the reader.

Modern scientific research is never isolated from the rest of society's activities, for there are numerous feedback loops that render science intimately symbiotic with the modern social order. Science depends on financial support from foundations and governments, and it also relies on the engineering community to develop improved methods of gathering and processing relevant data. Meterological research scientists, for instance, are deeply indebted to the engineers who developed the remote measurement techniques of Doppler radar and satellite telemetry, and today's seismologists have gained access to whole new vistas of inquiry through recent advances in microelectronic accelerometers and related sensing devices. Scientific research has progressed far beyond the point where it can depend upon human sensory observations alone. Because Mother Nature chooses to barely whisper her answers to our most compelling questions, science must turn to technology to extend the range and sensitivity of our feeble human senses.

Yet science is not technology. Science is a process of seeking answers to questions, some of which may turn out to be unanswerable, and some of whose answers may turn out to be irrelevant to the public. Technology has the more practical goal of reordering our natural environment to meet immediate human needs and desires. Sometimes the two activities overlap and reinforce each other, but sometimes they don't. No one knows how to determine in advance just what lines of scientific inquiry might eventually turn out to be useful to society. To do science is to take a chance on running down a blind alley, either discovering nothing, or else discovering only detritus that no one can use.

A recently abandoned research project in Japan, for instance, searched for a link between the behaviors of catfish and imminent earthquakes. The link indeed turned out to exist, but only sometimes, and this scientific finding was worthless to the engineers who still await the discovery of a reliable scientific principle on which an earthquake-forecasting technology might be built. But should the research on catfish never have been done? On the contrary, it needed to be done, because how else would we learn that this particular line of inquiry would be unproductive?

Science is always driven by ignorance, and it is in this spirit that I write this book. As society seeks to make sense of the seemingly random natural disasters that threaten human lives and labors, it appropriately asks what reasonable expectations it can have of the scientific community in helping predict and/or mitigate future catastrophes. I personally don't have the final or definitive response, nor does anyone else. What I offer instead is some (hopefully thoughtful) perspectives on a selection of historical natural disasters, the scientific progress that has been made in understanding them, the scientific challenges that remain, the socioeconomic factors that influence what scientific questions may be pursued in the future, and the prospects for achieving a level of scientific understanding that may someday permit us to predict, and ideally mitigate, natural disasters.

I write, not to disaster specialists, but rather to the broader community of professionals, policymakers, students, and independent thinkers, present and future, whose decisions or indecisions may profoundly affect the lives of those threatened by tomorrow's disasters. I write with the sincere hope that I can encourage a few readers to delete the phrase "common sense" from their lexicons and instead to critically evaluate all preconceived notions, including their own. Beyond this, I hope I also succeed in conveying some sense of the intellectual excitement in humankind's struggle to unravel Mother Nature's deepest secrets, through the fallible but self-correcting human activity we call "science."

1

Life on Earth's Crust

Lisbon's Longest Day

It began as a bright Saturday morning: November 1, 1755. It ended in death and destruction on a scale that would permanently upset the power balance of the Portuguese colonial system. The event was not a war, nor a political revolution, but a series of quite common natural phenomena – earth tremors and sea waves – that happened to be more energetic than usual. On the scale of Planet Earth, the incident was but a hiccup. On a human scale, it was a major natural disaster – one of the most devastating ever to strike a population center in the Western world.

At daybreak, Lisbon was home to 275,000 people, not counting the sailors and travelers one always finds in a busy port. A day later, those remaining numbered only in the hundreds: clerics administering last rites to the dying, looters scavenging valuables from the rubble of collapsed buildings, and some of the bolder survivors frantically searching for loved ones or trying to salvage fragments of their personal property before the flames got too close.

Hundreds of magnificent masonry buildings that had been built to stand for centuries (if not millennia) tumbled to the ground that dreadful morning, crushing and burying thousands. As some survivors fled to the surrounding hills, the less fortunate thronged to the wharves on the riverfront and were swept away by a series of tsunamis (seismic sea waves). Then came the fire. Exposure, injury, and starvation led to additional deaths in the following winter months. The overall death toll? No one knows. Estimates have ranged from 5,000 to 70,000 deaths in the immediate disaster[1,2]; the contemporary French philosopher Voltaire quoted a figure of 30,000.[3] Most modern references cite 30,000 to 40,000 fatalities in the first few days of November 1755, and approximately 20,000 additional deaths in the following months.

Population records at that time were kept by parishes, and, of the forty parish churches in Lisbon in 1755, twenty were completely destroyed and all were damaged. Many of the Lisbonites who hastily evacuated the city may have decided never to return, and, if they did return, some may not have been inclined to reaffiliate with their churches. Given the fire and the seismic sea waves, body counts were not practical, nor was such a gruesome accounting procedure high on anyone's list of priorities. Those who were curious about the mortality statistics had but one way to proceed: begin with the initial population figure, subtract the population that was still alive and present after the disaster, and attribute the difference to the grim reaper. The laws of mathematics give a precise figure. The laws of epistemology guarantee that the mathematically precise figure is wrong.

As for what actually happened in the Portuguese capital on that terrible day, in qualitative terms the documentation is a bit more consistent.[4] November 1 is All Saints' Day, a holy day of obligation for Catholics, and on that Saturday morning in 1755 all of Lisbon's churches were filled to capacity. Around 9:40 AM, the worshipers in the central cathedral were startled by the sound of thunder rising from the floor, the ominous rumbling quickly growing to a volume that drowned out the pipe organ and choir. The building itself trembled briefly in response, then everything and everyone fell silent. Some of those near the entrances dashed out into the streets just in time to witness the arrival of a series of three ground-heaving surface waves. Previously flat streets wiggled both horizontally and vertically, and distant observers described the city as swaying like a grainfield in the wind. Because masonry is not capable of withstanding significant bending stresses, the stone churches throughout the city promptly collapsed and entombed thousands of the faithful. Great clouds of fine dust rose from the rubble and shadowed the city from the sun. Through this billowing shroud came the wails of agony from the injured and dying. Most of the smaller houses and buildings that survived the first two earthquake waves succumbed to the third. Within a total time of little more than 3 minutes, a city that was one of the splendors of Europe lay largely flattened.

When people are threatened by powers beyond their retaliatory capacities, flight is a natural human reaction. Survivors of the earthquake, at least those who were capable, immediately fled. The lucky ones ran into the semicircle of hills around the city. The less fortunate chose the riverfront on the Tagus.

Lisbon is 12 kilometers (8 mi) upstream from the mouth of the Tagus River; in this stretch, the river's width averages around 3 kilometers (2 mi). At the city, the river widens further into a great inland bay, the Mar de Palha, some 11 kilometers (7 mi) wide. As early as 1200 B.C. the Phoenicians had recognized this place as an excellent spot to build a seaport, providing safe harbor from the onslaughts of an occasionally angry ocean without sacrificing easy ocean access. This unique geographical feature played no small part in Lisbon's eighteenth-century prosperity. It also, unfortunately, was a major contributing factor to the destruction of November 1, 1755.

No one expected the tsunamis. The first of the three great waves surged upriver from the ocean around 11 AM, some 80 minutes after the initial earthquake. It tore ships from their anchors, smashed them into each other, then dashed their wreckage into the masses that were now crowding the shore and quays. In their retreat, the waters swept away all the waterfront storehouses that had at least partially survived the earthquake and, along with these, a great mass of humanity. All published descriptions of the disaster describe with horror the arrival of these huge sea waves and their devastating effects.

A wide range of estimates can be found regarding the height of the tsunamis.[5] The best-informed opinion[6] seems to be that the three principal waves ranged in height from about 4.6 meters (15 ft) to 12.2 meters (40 ft), depending on which of the waves one speaks of and from what point on the riverfront one observed it. This would be consistent with reports of a 5.5-meter tsunami at Cadiz, some 350 kilometers to the south (Fig. 1.1). Note that a 12-meter wave crest is some four stories high!

But height alone doesn't tell the whole story of a tsunami. Tsunamis not only have great height when they strike shore, they also have a considerable wavelength, usually hundreds of kilometers long. A tsunami will pour in continuously for 15 to 30 minutes (and sometimes longer). Then, for the next 15 to 30 minutes, all of this water rushes back out. This is followed by the next wave crest, for a tsunami is never just a solitary wave. The bodies of those who perish in a tsunami are not likely to be found, for most will be dragged out to sea. In fact, history has provided us with relatively few eyewitness accounts of the world's major tsunamis, for the simple reason that very few eyewitnesses survived.

As devastating as the tsunamis were to the commercial and public buildings, the great waves did not extend their grasp to the higher ground, where most of the residential structures stood. By early afternoon, those Lisbonites who had survived both the earthquake and the sea waves might have figured that the worst of the ordeal had passed. If so, they were quite wrong. Within a few hours, the scattered small fires ignited by overturned stoves and lamps were whipped into a major conflagration by an onslaught of high winds. Fueled by the splintered remnants of the city's structures, the fire

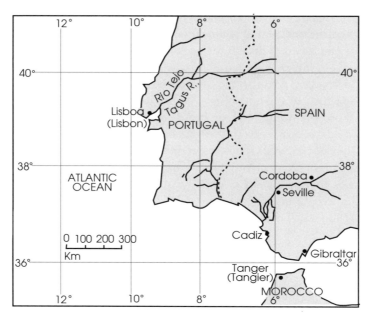

Figure 1.1. Geographical region affected by the earthquake and tsunamis of November 1, 1755.

completed the destruction of the city. A writer discussing the event in 1902 ventured the following assessment: "This fire, which lasted four days, was not altogether a misfortune. It consumed the thousands of corpses which would otherwise have tainted the air, adding pestilence to the other misfortunes of the survivors."[7] In fact, Lisbon had experienced an outbreak of plague as late as 1723, and the memories of many survivors went back that far. Yet one must wonder whether those buried alive in the rubble would have agreed that the fire was a blessing.

When the embers finally died out, survivors were confronted with the cruel fact that most of the city's food supplies had been burned or washed out to sea. Winter was approaching. Looting became rampant (attributed by some to the breach of the prisons during the initial earthquake), and murders were reported. In the days immediately following the disaster, a pound of bread is said to have been worth an ounce of gold. Although the palace in Lisbon was largely destroyed, the royal family had the good luck to have been at its country retreat in Belem at the time. On his return to Lisbon, 37-year-old King José ordered the state purchase of all available supplies of grain and ordered a system of relief distribution for food and medical supplies. Martial law was established on November 4, and within a few days thirty-four people were executed for looting. Responsibility for coordinating relief, peacekeeping, economic stabilization, and rebuilding fell to the king's minister Pombal, who used the opportunity to catapult himself into a twenty-two–year career as dictator.

The initial earthquake had been felt over the whole Iberian Peninsula, some of France and Italy, and parts of northern Africa,[8] and there were reports, probably apocryphal, that a town of 8,000 in Morocco had been completely swallowed up in a fissure.[9] The sea waves were recorded in England, Ireland, and the West Indies (though of course they no longer posed much of a threat after traveling such distances). The intensity of the earthquake has since been estimated as 8.75 on the Richter scale; if accurate, this would make it one of the three or four strongest earthquakes experienced on the planet in the last two hundred fifty years.

Many of Lisbon's survivors lived for the next year or two in tent cities set up in the hills. Most were in no special hurry to return to the city's ruins, for there were persistent aftershocks: 30 in the first week after the disaster, a violent shock on November 8, and a sizable earthquake on December 11. In August of 1756, it was reported that there had been 500 aftershocks since the previous first of November. Of course, none of this is quantifiable according to modern scientific standards. The science of seismology and its supporting instrumentation did not emerge until the late nineteenth century.

Lisbon was eventually rebuilt by drawing on the wealth of the Portuguese colonies (most notably Brazil) and through the foreign aid that poured in (principally from England). Yet many cultural treasures were lost forever. Hundreds of paintings by the masters, including works by Titian, Rubens, and Correggio, were reduced to ashes. Collections of original maps and charts relating to the early Portuguese voyages of discovery were destroyed. A thousand or more hand-written manuscripts, including a history written by Emperor Charles V in his own hand, were gone. All of the public and private libraries, with hundreds of thousands of early works and archival documents, had been completely consumed. In one fell swoop, Portugal irretrievably lost a major portion of its cultural heritage.

Today, few would think of the Lisbon earthquake when asked to name a major natural disaster. The event has long faded into obscurity, relegated to an occasional list in an almanac, which may even get the date wrong.[10] On a human time scale, one does not immediately relate to something that happened in 1755. On a geophysical time scale, however, two hundred forty

years is not a long time at all. What happened at Lisbon is eminently relevant to our understanding of the human species' tenuous relationship with
its restless planet.

The Demise of an Island Civilization

Some 220 kilometers (140 mi) southeast of Athens and 110 kilometers (70
mi) north of Crete, a beautiful little island graces the Aegean Sea. From the
air, one sees the unmistakable crescent shape that is apparent on maps (Fig.
1.2). In earlier times this place was called "Thera"; today it is more commonly called "Santorin," or "Santorini." The 75 square kilometers (30 mi²)
that jut from the sea represent a little less than half of the island's original
surface area in 1626 B.C. The part that's missing, down to a depth of over 300
meters (1,000 ft) below sea level, was blown into the atmosphere in a cataclysmic volcanic explosion whose precise date is still the subject of debate
between physical scientists and archaeologists. In quoting the date of 1626
B.C., I'm siding with the physical scientists (Table 1.1).[11]

Figure 1.2. The Aegean islands of Santorini (Thera) and Crete.

Table 1.1. *Dating the Bronze Age explosion of the Thera volcano*

Date (B.C.)	Uncertainty (yrs)	Technique	Samples
1640	30	Carbon-14	Trees buried in initial ashfall
1626	1	Tree rings	Bristlecone pines in USA (suppressed growth)
1626	1	Tree rings	Oak in Irish peat bogs (overlapping samples)
1643	20	Chemical	Ice cores in Greenland ice sheet (acid snow)
1617	~20	Literary	Ancient Chinese manuscripts
1500	~50	Art styles	Minoan pottery
1500	~50	Art styles	Egyptian pottery and tomb paintings

Source: See the sources cited in note 11 to Chapter 1.

Uncertainties are an inherent characteristic of all scientific data. This does not, however, mean that such uncertainties are just a vague hand-waving description; in practice, even estimates of uncertainty must be substantiated through rigorous analytical standards before any scholarly publication will accept them. Nevertheless, it comes as no surprise to practicing scientists that data originating from different techniques applied to different samples are found to disagree. The surprise comes when they *agree*. And when that happens, a scientist begins to suspect that there is some element of fundamental truth underlying all of the abstract mathematics and data tables.

Today no archaeologist doubts the essential details of the event: A remarkably modern Bronze Age city, with running water, multistory residences, and (judging by its art) social standards of gender equity, was buried in volcanic ash that preserved its wonders for rediscovery thirty-six centuries later. At the same time, more than half of the rest of this little island was blown into earth's atmosphere, and the sea rushed in to fill the void. Much remains to be discovered here, but currently the main archaeological debate is whether Thera's cataclysmic volcanic explosion of about 1600 B.C. can be linked to a more profound and puzzling event: the subsequent demise of the entire Minoan civilization that then thrived on the northward-facing shores of Crete. Many writers (myself included) are convinced that there is indeed a substantive link here. Yet, had Thera's catastrophe occurred a century later, it would be a great deal easier today to develop irrefutable arguments in support of this view. Scientists, after all, are in the business of being skeptical, and many today remain skeptical that Thera's Bronze Age explosion had consequences that extended far beyond this little island. None doubt, however, that the explosion did indeed destroy the local civilization.

The present city of Thera, with its immaculately whitewashed buildings and domes, is built right up to the edge of a 300-meter (1,000-ft) cliff that drops nearly vertically to the sea. There is no way to take a bad photograph of Thera, and one sees various scenes of this town over and over on Greek travel posters. Although many of Thera's buildings are centuries old, there is no evidence of anything truly ancient in the town itself. Until just fifteen years ago, the only way to get between the city and the sea was to ride a donkey down a stepped and switchbacked path that had been carved into the cliff. Today the less adventurous can ride a modern funicular.

Cruise ships stopping here cannot drop anchor, because of the great depth of the water, and special moorings have been built to accommodate them. Near the middle of the deep blue bay is the small and virtually lifeless volcanic island Nea Kameni (or "New Burnt"). Although this volcano juts only 30 meters (100 ft) or so above the water that surrounds it, it has actually rebuilt itself 450 meters (1,500 ft) from the sea floor during the three millennia since it exploded so catastrophically. Curious visitors can easily get to the volcano by boat, tramp around the few square kilometers of sharp lava, and peer into the half dozen or so vents that still almost continuously spew sulfur-laden vapor. Standing on Nea Kameni and looking back toward the city, one gets a dramatic view of the cliff that was sheared from the now-vanished part of the island in that cataclysmic explosion some thirty-six centuries ago. Clearly, this was not the place to be in 1626 B.C.

The southern and eastern shores of the main island have several black beaches, similar to those found on some eastern shores of the island of Hawaii. The material on these beaches has the approximate consistency of sand but is in fact volcanic lava that was pulverized by its own internal stresses when it was rapidly cooled by contact with sea water, then was further ground up by wave action. In the immediate area are dozens of small but thriving vineyards and family-run wineries. All water must be drawn from deep wells; the deep volcanic soil is too porous to permit the accumulation of surface water.

In the southern part of the island, near the town of Akrotiri, is a remarkable archaeological excavation: several hectares under fiberglass roofing, where dedicated researchers have been working since 1967, a spoonful at a time, to retrieve evidence of Thera's predisaster civilization (Fig. 1.3). Most of the 20 meters (65 ft) of volcanic ash that once buried the site was eroded by natural forces over the centuries, and occasional artifacts began to peek out from the soil by the 1940s. The organized dig has so far unearthed the equivalent of about two blocks of a city whose total diameter was at least 1.5 kilometers (1 mi). The only structures uncovered so far are residences; we know nothing about the public buildings or palaces (assuming there were any). At the present rate of excavation, there are at least three hundred years of archaeological work left to do here.

Greece's former inspector general of antiquities, Spyridon Marinatos, began studying evidence of Thera's ancient catastrophe in 1932 and continued (on and off) until he died at his Akrotiri excavations in 1988.[12] He is buried in the ruins of one of the ancient buildings. The Bronze Age city he discovered is not in the history books, and no written records have been found that even hint at its original name. It is now sometimes referred to as Akrotiri, after the village just upslope from it. More often the current literature refers to "Thera," the older name for the island and the current name of the island's principal city.

One thing that is apparent: Ancient Akrotiri was magnificent! It had running water a thousand years before any other city we know of. Some of this

Figure 1.3. A view of the archaeological excavations at Akrotiri.

water ran in open stone culverts through the city, but some also ran through residences that had interior toilets and baths. (A presently unconfirmed speculation is that some of these streams carried hot water drawn from geothermal sources.) Residences were two, three, and four stories high. Their construction shows evidence of at least some understanding of earthquake engineering: Walls tend to meet at oblique rather than right angles, and lintels and doorframes were made of wood rather than masonry (wood withstands dynamic tension loading far better than does stone). The buildings are full of pottery, skillfully executed. Bronze tools have been found, and even a piece of glass that appears to be a lens.

Most relevant, however, is the art. Each private dwelling unearthed to date has multiple interior walls covered with elaborate frescoes whose quality rivals the best *public* art found in other ancient excavations.[13] The flora and fauna depicted in these frescoes make it apparent that this early civilization traded with northern and eastern Africa. Women and men are represented as equals. Some ships are shown (Fig. 1.4), but these appear to be trading vessels rather than warships. In fact, not a single one of the frescoes unearthed shows a theme that is even remotely militaristic, political, or jin-

Figure 1.4. One of many Bronze Age murals discovered at Akrotiri. (Photo courtesy National Archaeological Museum, Athens.)

goistic. These findings stand in dramatic contrast to the majority of the art that has been excavated at other ancient sites dating from the following two millennia.

It must be remembered that individuals don't invest their resources in art until their more essential human needs have been met. Yes, repressive governments sometimes do commission public art while their people lack basic needs, but when this is done, such art depicts themes that support the currently state-endorsed political philosophy. At Akrotiri the archaeologists have found wall-sized private art, lots of it, but *none* reflecting themes of political or military power. And, at least so far, they've found no public art at all.

The conclusion is compelling that these were a sophisticated, prosperous, egalitarian, and peaceful people. (It does not follow, of course, that they were necessarily democratic; they may in fact have been ruled by Plato's mythical perfect autocrat, the philosopher-king.) Meanwhile, the art and few remnants of recovered script also make it clear that ancient Akrotiri and contemporary Crete had a shared language and closely related cultures. With favorable winds, one could travel from Thera to Crete in less than a day, even in ancient times. And the inhabitants of Akrotiri seem to have been accomplished seafarers, as were their Minoan cousins on the larger island of Crete.

The excavations at Akrotiri have uncovered a few animal skeletons, but (as of this writing) no remains of human victims and virtually no coins, jewelry, or other easily portable valuables. This suggests that the volcano gave ample warning, and that the populace was smart enough to heed it. In fact, the excavated ashfalls suggest that as much as a year or two may have passed between the abandonment of Akrotiri and the cataclysmic volcanic explosion that buried the city for the next thirty-six centuries.

But where did these people go? Although there is no hard evidence, it seems likely that most would have gone to Crete, at least initially. Biblical references suggest that some may also have migrated to Palestine (the Philistines of Judaic Scripture were said to have come from "Caphtor," or Crete). Correlations among other ancient documents support speculation that some of the émigrés resettled on the western coast of Italy and on the northern coast of Africa. But regardless of where they went, the sad fact remains that the splendor of their homeland was never duplicated elsewhere in the following decades. More immediate human needs take priority over the development of technology and art. And if the artisans get old and die before getting a chance to pass on their specialized knowledge, a whole society will take a big step backward.

Why did they believe they had to leave, and how did they all manage to agree about this? With the benefit of hindsight, today we can acknowledge that evacuation was the correct decision. But given the messy dynamics of human decision making, what compelled the ancient Akrotirians to agree at that time? Given the lack of cultural evidence of a strong police force or army, it's unlikely that the evacuation was imposed by martial law. On the other hand, it's also unlikely that Akrotiri was the only settlement on the island of Thera about 1630 B.C.; it could easily have been but one of a small nation of cities on the island at that time. The level of commerce represented in the excavated frescoes is simply not consistent with the size of the single city that has been discovered. In fact, a few human skeletons with artifacts dating to about 1600 B.C. have been found buried in volcanic ash at the northern end of the island, where there still have been no organized excavations. Will ancient cities be found here as well? Quite possibly they will. (As for the the half of the island that was blown into the atmosphere, any evidence of human activity there was irretrievably vaporized thirty-six centuries ago.)

Thera's ancient volcano jutted some 1,500 meters (5,000 ft) above the sea (Fig. 1.5), but it was probably centered quite near the present site of Nea Kameni. And if Nea Kameni were to explode while you were visiting this island today, your chances of survival would be best if you were – that's right – sipping a beer at the archaeological site at Akrotiri!

A reasonable conjecture, then, is this: Ancient Akrotiri was but one of many cities on the island of Thera in 1628 B.C. Because the volcano had been dormant for two millennia,[14] the inhabitants did not consider it a threat (much as the residents of Naples do not consider Vesuvius to be a threat

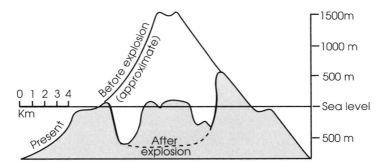

Figure 1.5. Cross-section of Thera running from southwest to northeast. Note that the horizontal scale has been compressed. (Adapted from J. V. Luce, *The end of Atlantis* [Athens: Efstathiadis & Sons, 1982].)

today). When the volcano reawakened, its initial blasts and ashfalls first devastated the most vulnerable communities in the northern and western parts
of the island. This shocked the Akrotirians into taking the volcano seriously,
and they evacuated – probably fully expecting to return someday. Had the
volcano not blown its top in 1626 B.C., they would have returned, shoveled
out their city, and continued their cultural and technological advances.

The Atlantis Legend

Most of what was written in antiquity has disappeared – some written
records succumbing to natural forces of decay, others to disasters, and a
good bit being destroyed purposely (as when the huge library at Alexandria
was burned by a mob in A.D. 415). Only a few remnants of ancient texts exist
today, and many of these survive only in translation. Some of these fragments tell stories about past catastrophes that compel us to ask questions
whose answers will forever remain incomplete.

The Greek philosopher Plato, writing around 360 B.C., described an
island he called "Atlantis."[15] According to oral history that preceded him by
a thousand years, this great seafaring civilization lay in the Atlantic Ocean
beyond Gibraltar. Atlantis was said to have disappeared beneath the sea in a
cataclysm, in a "single day and night of misfortune," leaving not a trace. The
possibility that Thera equates to Atlantis has been examined by quite a few
writers,[16] with varying degrees of credibility, for at least the past fifty years.
The nagging difficulty is that Plato seems to have placed Atlantis in the
Atlantic Ocean, not in the Aegean Sea. Could the geography have gotten
befuddled to that degree in an age when the known world was so small?
Although this is not an easy issue to resolve, many of the circumstantial similarities between Plato's descriptions of Atlantis and the recent discoveries at
Akrotiri make one wonder . . . Though all of Thera didn't sink beneath the
sea, at least half of it did indeed vanish. Perhaps this was the grandest half,
and the shock of the discovery of its disappearance may have led ancient seafarers to weave the accounts that ultimately inspired Plato's writings.

Alternatively, one might take Plato literally and place ancient Atlantis in
the Atlantic Ocean a few hundred miles off the coast of Lisbon. This leads
to the seductive idea that Lisbon's great disaster of 1755 may have been precipitated by rumblings from Atlantis's watery grave. To date, however, this
speculation is contradicted by the geophysical mappings that have been con-
ducted of the floor of the Atlantic Ocean. Although there is indeed evidence

that volcanic mountains have sunk beneath the sea here, they seem to have done so very slowly, over the course of many millions of years. Volcanic islands do not subside on human time scales; when such islands do disappear suddenly, it is because most of their volume explodes into the atmosphere.

There is, however, something we *can* conclude from Plato's accounts of Atlantis: The threat of natural disaster has occupied the minds of thinkers for a very, very long time.

The Extinction of the Minoan Culture

Let's return to the possible link between the Bronze Age explosion at Thera and the archaeological consensus that the entire Minoan culture disappeared within the following one hundred fifty years. In 1650 B.C., the island of Crete was home to some 250,000 Minoans, around 40,000 of whom lived in the capital city of Knossos (whose partially reconstructed ruins can be visited today). Bronze Age Crete was a great nation of seafarers, who dominated the Mediterranean sea trade and may have ventured as far as England. Minoan culture was highly respected by the contemporary Egyptians at a time when Egyptians considered other foreign nations to be inhabited by barbarians. The Greek mythological tales of Theseus and the Minotaur, and of Icarus and Daedalus, are clearly set in Minoan Crete (although the stories were not written down until many centuries later).

Then, within just a few generations, in the sixteenth century B.C., the Minoan language was replaced by a significantly different language, the art changed, the public buildings fell into ruin, and maritime commerce ended. Mycenaeans from mainland Greece moved into Crete, bringing their own customs and artistic styles with them. And yet no evidence of hostilities has ever been found. The thriving Minoan culture seems to have simply disappeared, replaced by a Mycenaean culture that was in no respect more advanced (and in some ways less advanced).

It should be remembered that entire cultures do not die easily. The Jews maintained their culture quite effectively through a period of two millennia without a homeland. The Poles had no country and their language and culture were suppressed for one hundred fifty years prior to World War I, yet they emerged as Polish as ever. King Harold II lost England to William the Conqueror in A.D. 1066, but the invaders became more English than the conquered English became Norman. Native Americans today maintain most elements of their original culture after centuries of conflict with an influx of

Europeans; in fact, in many pockets of Latin America the Native American culture still remains dominant.

So, what happened to the Minoans? It's fairly clear they weren't killed off in a war. If there was an epidemic, we have no record of it. On the other hand, we *do* know that a few generations earlier there was a cataclysmic explosion at Thera, just 110 kilometers (70 mi) to the north. And we also know what havoc the explosion of a sea-level volcano can wreak on neighboring shores.

In 1883, the volcano Krakatau in the East Indies exploded, tragically killing some 36,000 and giving Victorian scientists an opportunity to make numerous in-depth studies of the event. Although we will return to Krakatau in some detail in a later chapter, I mention it here because our understanding of the Krakatau eruption helps us to reconstruct what happened after Thera's Bronze Age explosion. These were very similar events: the explosion of a sea-level volcano that ejects great masses of material into the atmosphere, then a collapse of the volcano walls, the sea rushing in to fill the void, and finally a great sea wave referred to as a "tsunami."

The volume of dirt and rock blown into the atmosphere at Thera was some 4 to 10 times that of Krakatau (we learn this from soundings made of the sea depths in the craters at the two sites). The initial blast surely killed any living creature that remained on either island. As the sea poured into Thera's newly formed crater, the current and turbulence probably overwhelmed any nearby Bronze Age boats. On the shores of Crete, 110 kilometers to the south, few would have paid attention to the initial rumblings – they probably sounded like no more than distant thunder. All, however, certainly jumped at the sound of Thera's final explosion (we know that Krakatau's final explosion was heard some 3,000 kilometers [2,000 mi] away). Around 15 minutes later (based on mathematical reconstructions of the event), the sea abruptly receded from the harbors and shores of northern Crete, exposing sea bottom as far as the horizon. Then, after another 15 or 20 minutes, the waters returned with a vengeance, at a speed of around 300 kilometers per hour (200 mi/h) and a height variously estimated at 30 to 90 meters (100–300 ft).[17] Some scientists have suggested that these waves may even have reached as high as 200 meters (600–700 ft)! But even assuming the lower figures, a series of these great waves surely rolled inland at least a few kilometers, far enough to engulf all structures in any way related to maritime commerce. Then, as these great waves receded (and remember, this takes 15 to 30 minutes for each wave), they swept away most of the wreckage. In fact, in archaeological excavations at Amnisos, the harbor town serving ancient Knossos, huge stone blocks have been found torn from their foundations and strewn toward the sea, consistent with what one expects as a great tsunami recedes.

The dynamics of such powerful tsunamis are certainly a matter of scientific dispute. The problem is that the mathematical formulas have been verified only for tsunamis with heights up to a few tens of meters, so perhaps such formulas can no longer be believed when they predict wave heights many times higher than has ever been observed. On the other hand, all we really need to be sure of here is a series of 40-meter (130 ft) waves. When Thera's volcano quaked in 1956 (and nothing at all was ejected from below sea level), tsunamis as high as 36 meters (120 ft) were reported at Amorgos and Astypalaea, roughly 80 kilometers (50 mi) to the northeast. As we saw previously when considering the Lisbon disaster, even smaller waves are quite sufficient to destroy wharves, warehouses, shipbuilding facilities, and other coastal structures that support maritime commerce. More serious, however, is the humanity such waves would have swept off the northward-facing shores of Crete in 1626 B.C. After all, who is more likely to live near the sea than those who derive their livelihoods from the sea? Shipwrights, sailmakers, sea merchants, sailors and navigators, mapmakers, and smiths all would have naturally resided in the areas most vulnerable to tsunamis.

It has been argued that the loss of whole fleets of ships, and even of wharves, should not have seriously affected the entire Minoan civilization on Crete. After all, these were the days when wooden ships were expected to have a short working life, and, tsunami or not, ships would be replaced routinely every decade or so. My personal response is that ships and physical infrastructure would have been far from the worst of the loss. In times when maritime crafts and technologies were passed on through apprenticeships rather than published handbooks, the loss of collective knowledge would have been a devastating blow that Minoan Crete simply could not recover from. There was no way to build new fleets, because everyone who knew how to do this had been drowned in the tsunami. Yes, the ships that survived the great waves (by virtue of having been in deeper water at the time) would surely have continued to ply the seas for the next few decades before succumbing to rot. At the end of their useful life, however, there would have been no way to replace them with new ships of comparable seaworthiness. The loss of knowledge is always more devastating to a culture than the loss of physical entities.

It seems unlikely, therefore, that Minoan maritime commerce would have ended abruptly with a great tsunami in 1626 B.C. Rather, the effect of such a catastrophe had to be a gradual decline that extended over several decades.

Eventually, the Minoan fleets deteriorated in quality and quantity to a level where they could no longer intimidate the Mycenaeans, and these mainland Greeks were finally able to land on the shores of Crete.

Normally, one would expect such a landing by alien peoples to precipitate a war. In this case there seems to have been more. Once again we can find an explanation in the 1626 B.C. volcanic eruption on Thera.

Akrotiri was originally buried in 20 meters (65 ft) of volcanic ash. Clearly, this ashfall did not end at the perimeter of the island. Some ash was blown into the stratosphere, where it affected global climate and resulted in the acid snow whose traces in glaciers permit us to date the event (see Table 1.1). Heavier components of this ash fell to the sea and earth. These also have been sampled, allowing scientists to reconstruct the fact that most of Crete and parts of Turkey were covered with Thera's ashfall. It takes very little ash (a millimeter will do) to halt photosynthesis in a plant leaf. A few centimeters will kill ground vegetation. Only a little more will poison the soil for years, until the acid leaches away. A few tens of centimeters (less than 1 ft) makes it impossible to grow any food crops for several decades. Wait a century or two, provide enough water, and the volcanic soil will be a farmer's best friend. But don't buy farmland under a recent volcanic ashfall and expect it to be profitable in your lifetime.

The necessity of agriculture is a given. Crete already supported a population of around a quarter million people when the refugees from Thera arrived around 1628 B.C. Then, a year or two later, came the simultaneous devastation of both the agricultural base and the maritime commerce infrastructure. We will probably never know any of the details of the bleak years that followed; all we know for sure is that the Minoan culture disappeared, replaced by the Mycenaean. Yet the evidence is fairly compelling that the collapse of this entire thriving culture resulted from a single geophysical event: the explosive eruption of an offshore volcano.

Can such an event happen again? The sad answer is yes. It not only can, but it probably will.

Of Life Spans and Disasters

The crust of our planet Earth is just that – a thin layer of solid material floating on a viscous planetary interior. Patches of crust constantly heave and drift in response to subterranean fluid forces, and nowhere on the globe is the solid surface truly stable. On the scale of the planet, the seas are puddles prone to splashing when the crust shifts. As Earth spins on its axis, its equator racing to the east at more than 1,600 kilometers per hour, the atmosphere swirls in great eddies and vortices, redistributing moisture through evaporation and condensation, and occasionally swirling about too fast for the structures we build to shelter our human activities. Meanwhile Earth swarms with life, most of it microscopic and some of it with an appetite for consuming the living machinery of the human body. Occasionally one of these microbial species finds a particularly efficient method of traveling from person to person and its population explodes, to the detriment of large segments of humanity. If this isn't enough to worry about, Earth's orbit through space crosses the orbits of thousands of asteroids, many of them large enough to wreak considerable havoc if they hit the planet (as the geological record suggests has happened on numerous occasions in the past).

Yet the human species has survived and prospered in the face of all these threats. The very existence of modern men and women can be credited to the disparity between two natural time scales: one the average period between major disasters, the other the span of the human generation. On an astronomical time scale, we produce progeny in a blink of an eye after we are born. And then, biologically worn out, we die and return our chemical components to the environment to be recycled in different and often more complex configurations. Although large meteor impacts may produce an occasional mass extinction – killing off whole species – such events occur at average intervals of many millions of years. Even the slow grind of evolution has had ample time to create biological marvels in the eons since the last major asteroid impact.

With earthquakes and volcanoes, the intervals between events are considerably shorter. Still, even seismically active regions seldom experience more that one major earthquake in a human lifetime, and volcanoes frequently lie dormant for several centuries between major eruptions. In most regions, it is unusual for life-threatening storms or floods to occur more frequently (on average) than once per human generation. Historically, mass epidemics have occurred at time intervals of around two or three generations.

The existence of the human species today is testimony to the statistical probabilities reflected in these relative time scales: The probability of growing to puberty, reproducing, and migrating must exceed the probability of premature death, if the species is to survive. If humans lived to 1,000 years and reached puberty at 300, the closer match between our life cycles and the frequency of the cataclysms of nature would have ensured our species' evolutionary demise long before we began using tools. No doubt this is why smaller life forms have shorter reproductive cycles; an ant, for instance, stands a good chance of being squashed or swept away in a trickle if it lives very long. If an organism delays too long in reproducing, relative to the threats it faces in surviving, nature claims it before it passes on its genes. Our life spans are well matched to the average intervals between the major upheavals of nature. Most of us are likely to die of other causes long before our bones are disturbed by a natural disaster.

Notice, however, that I speak in terms of probabilities here. Given enough time, enough geographical area, and enough people, individually improbable events become increasingly likely to happen. In the world as a whole, we are guaranteed to hear about a few natural disasters doing great damage and claiming human victims *within the coming year.* The probability of our individually being counted among the victims is small, but the probability of there being numerous other human victims is 100%.

As humans, we care about more than statistics; we care about individuals. And we care most about our loved ones. Who among us hasn't watched the scene of some disaster reported on the news programs and wondered how effectively we personally would handle such adversity? Who among us hasn't been awed by descriptions of the fury unleashed by killer floods, tornadoes, hurricanes, earthquakes, volcanoes, avalanches, and the like, and wondered about the individual victims? And who among us hasn't been drawn to the human interest stories of those who managed to survive great disasters while those around them perished? We know, at least subconsciously, that none of us is immune to the threats of nature unleashed, and we find comfort that there can be hope in even the worst. It's part of our human nature to choose life over death.

To keep the scope of this book manageable, I have tried to maintain a distinction between natural disasters and sociopolitical or other man-made disasters. Clearly, there are areas of overlap. Wars usually claim more victims through disease and deprivation than through bullets; Napoleon, some writers have claimed, lost as many soldiers to measles as to any other single cause. Yet in most cases my distinction is valid, in that it identifies the class of events whose explanations have some chance of yielding to the methods of empirical scientific inquiry. To this end, I will use the following definition:

> A *natural disaster* is an event in which the forces of nature claim human lives or destroy the fruits of human labor on a large scale.

Although, for instance, bridges have been known to collapse when their piers were undermined by raging streams, the scale of such an event does not qualify it as a "disaster" (even though the engineering professors certainly study each wreckage long past the attention span of most of the public). On the other hand, if a fallen bridge is one of many structural failures associated with a major storm, the storm is considered the disaster, and the bridge failure is one of many details that we will legitimately incorporate into an overall picture of the storm's effects on the affected region's infrastructure.

The scale of the event is also relevant for a more pragmatic reason: Large-scale events attract widespread attention, and this in turn attracts funds for scientific investigation. As a result, we tend to *learn* more about killer hurricanes than about the lightning bolt that killed Cousin Charlie on the local golf course. Although Cousin Charlie's demise was no less a tragedy for his loved ones, the scale of such an event falls considerably short of attracting a swarm of scientific researchers.

Yet, scale notwithstanding, the human connection remains fundamental. Very little science is done as a search for objective truth independent of human concerns. Science is a human activity, supported by society through its institutions, with the expectation that the new knowledge generated will, on average, lead to positive social and economic consequences. We crave to know most that which affects us most. Thus do we define, in a universe of possible questions, the lines of inquiry we will reward our scientists for following. If earthquakes occurred only in Antarctica, there would be considerably less research on earthquake prediction.

As a field of scientific inquiry, natural disasters present monumental challenges to those who study them. I devote most of this book to discussions of these various challenges, so here I'll mention only one: irreproducibility. If you are researching the properties of a new organic chemical compound, you can easily check yourself by repeating your laboratory tests. In fact, if you are a good scientist, you repeat the tests many times with many samples. On the other hand, a phenomenon like a volcanic eruption gives you only one shot. You can't re-create the volcano in the laboratory, much less test it repeatedly. For additional data, you need to wait for a new volcanic eruption, which is certain to differ in fundamental ways from the first, and you need to be lucky enough to have your instruments in the right place at the right time. Knowledge in a field like this is not likely to race forward at breakneck speed.

To a great extent, and more than most of the public may realize, all science rests on historical foundations. No science student is expected to repeat Pasteur's experiments in developing a vaccine for rabies or Fizeau's experiments in measuring the speed of light in moving fluids. Scientists acknowledge these results on the weight of their historical documentation – accepting the fact that the experiments were actually performed in the past, that they yielded the reported results, and that the methodologies have already been critically reviewed and analyzed. No student of science has the time to reconstruct all of today's scientific knowledge through firsthand experimentation. Instead, students are assigned textbooks that reflect consensus among the intellectual leaders in the discipline, whose expert judgments are in turn based on historical experiences or documents. What science students learn is largely history.

I mention this to dispel any notion that a dependence on historical sources might in some sense be "unscientific." Digging through historical sources to gain an understanding of natural disasters is entirely consistent with standard practice in the sciences. Of course, it's also true that in studying natural disasters we have little other choice. Because major disasters occur at extended time intervals, historical sources are our only way of linking individual events into a *class* of phenomena. A description of a single event is not science. Single events are relevant scientifically only insofar as they give us clues to the larger patterns in nature.

The challenge of disaster science, then, is to identify patterns in classes of events that are not only geographically dispersed, but whose time lines also transcend the life span of the individual observer. The irony is that our species could not have emerged and survived on this turbulent planet to ask such questions today if natural selection hadn't endowed us with a short average life span relative to the major upheavals of nature. Yet because we live so few years, a single scientist cannot make enough firsthand observations in a lifetime to develop anything approaching a predictive science of natural disasters. Our scientific understanding can progress only through the cultural legacies of other humans, long deceased, and through the interdisciplinary collaboration of thousands of living scientists who individually have had time to learn just a little bit new in their short tenure on this turbulent planet. In the chapters that follow, I share with the reader an overview of our current understanding of natural disasters that reflects such collaborative and interdisciplinary intellectual efforts, and I point out a few of the many problems and questions that continue to haunt us today about natural disasters.

Notes

[1] Estimated by J. J. Moreira de Mendonca in *Historica Universal dos Teremotos . . . com uma narracam individual do Terremoto do primeiro de*

Novembro de 1755 . . . em Lisboa (Lisbon, 1758). This author, however, may have had reason to underestimate casualties to soothe possible concerns of governors of distant Portuguese colonies.

2 Jose de Oliveira Trovao e Sousa, *Carta em que hum amigo da noticia a outro do lamentavel successo de Lisboa* (Coimbra, Dec. 20, 1755), pamphlet. Other inaccuracies in this publication raise questions about the credibility of the casualty estimate of 70,000. This author's theme was that Lisbon was punished for its wickedness, and to buttress this argument he may have overestimated the death toll.

3 Voltaire's satire *Candide* was first published in 1759. Candide washes ashore after a shipwreck just in time to witness the destruction of Lisbon, then is arrested and flogged by officers of the Inquisition, who are looking for scapegoats to explain why God has chosen to punish the city. Candide's philosopher-mentor and traveling companion is hanged for heresy for suggesting that the event had a natural rather than a supernatural cause. Officers of the Inquisition, their court reduced to rubble, expected such decisive actions to prevent future earthquakes.

4 Most of what follows here is based on T. D. Kendrick, *The Lisbon earthquake* (Philadelphia: Lippincott, 1956), which remains the best comprehensive English-language source regarding the 1755 disaster.

5 Kendrick, *Lisbon earthquake,* gives 15 to 20 ft, whereas a figure of 50 ft appears in C. Morris, *The destruction of St. Pierre and St. Vincent and the world's greatest disasters* . . . (Philadelphia: American Book and Bible House, 1902). The latter author may tend to exaggerate, for he also repeats a story, discredited by Portuguese engineers soon after the disaster, of a quay being swallowed up in a fissure and taking a number of moored vessels and a great crowd of humans down beneath the waves with not so much as a splinter or shred of clothing floating back to the surface. The account goes on to say that subsequent measurements revealed a depth of almost 100 fathoms (180 m, or 600 ft) at this spot. There appears to be no record of any such measurement having been made at the time. Disasters frequently precipitate such unsubstantiated "data."

6 T. S. Murty, *Seismic sea waves; Tsunamis,* Fisheries and Marine Service, bulletin no. 198 (Ottawa, Can.: Fisheries and Marine Service, 1977).

7 Morris, *Destruction,* 408.

8 H. F. Reid, The Lisbon earthquake of November 1, 1755, *Bulletin of the Seismological Society of America,* 4 (2) (June 1914), 53–80.

9 This claim can be found in several secondary sources, but none of the authors identifies the primary source (assuming there was one), nor does anyone give the name or precise location of the Moroccan town. The story may say more about human nature than about geophysical phenomena.

[10] Numerous publications erroneously list the date of the Lisbon disaster as November 1, 1775, rather than November 1, 1755.

[11] M. K. Hughes, Ice layer dating of eruption at Santorini (Thera), *Nature,* 335 (1988), 211–12; C. U. Hammer, H. B. Clausen, & W. F. Friedrich, The Minoan eruption of Santorini in Greece dated to 1645 B.C., *Nature,* 328 (1987), 517–19.

[12] S. Marinatos, On the chronological sequence of Thera's catastrophes, *Acta* (1971), 403–6; Marinatos, Thera: Key to the riddle of Minos, *National Geographic,* June 1972, 702–6.

[13] Many of the restored frescoes are on display at the National Archaeological Museum in Athens.

[14] There seems to be another city beneath the Akrotiri excavation, also buried by a volcanic ashfall, but two thousand years earlier.

[15] Plato gives a short account of Atlantis in his *Timaeus* and a longer account in *Critias.*

[16] A particularly well-researched and persuasive argument is presented by J. V. Luce in *The End of Atlantis* (Athens: Efstathiadis & Sons, 1982).

[17] Of course, there are no eyewitness accounts to tell us the heights of these waves. The quoted estimates are based on mathematical models in which one considers the amount of geophysical energy released, the degree to which this energy is likely to have been coupled into the sea water, the profile of the sea floor, and the equations which are known to describe the dynamics of sea waves. There are many sources of uncertainty in such calculations. See Murty, *Seismic Sea Waves,* for a summary of the mathematical theory.

2

The Evolution of Science

Newton's Clockwork Universe

In 1666 an outbreak of bubonic plague struck Cambridge, England, and the administrators of that town's university wisely closed its doors for a year. Young Isaac Newton (1642–1727) went home to his family's farm, sat beneath a tree, and gazed at the moon. Maybe an apple fell from the tree, and maybe it didn't. No matter, for since prehistoric times it had been no secret that things fall – whether apples from trees, arrows in flight, or bison stampeding over a cliff. But Newton's mind wasn't on apples, arrows, or bison; he was thinking about the moon that hovered over his head. Why wasn't the moon falling from the sky, just like everything else? Could it be that maybe it *was* falling, and that it just didn't get any closer because of the peculiar nature of its motion as it fell? Hmmm . . .

What happened next is a very long story that has been retold and embellished in many forms, supplying the material for hundreds of different books over the last three centuries. In brief, young Newton succeeded in formulating no less than a theory of the mechanics of the entire universe. Newton's laws encompass the laws that Johannes Kepler had formulated earlier for the special cases of the planets and Galileo's laws for accelerated motion, but they also go much farther, predicting the tides, the orbits of the moons of the planets, the motions of windmills, and the forces developed in the individual pieces of complicated structures. These predictions are numerical and can be tested through measurement. Newton's equations work so well that we routinely use them today in ways Newton never would have dreamed of: in automotive design, in navigating space probes, and in helping analyze geophysical and meteorological phenomena.

The linchpin to this success story was an insight of Newton's that can be paraphrased as follows:

> The laws of nature are the same at all times and at all places in the world and in the Universe as a whole. The details of specific events will be different, but the underlying principles will always be the same.

Of course, this can't be proven, and no one has ever proposed a good reason to explain *why* the universe must behave in such a consistent fashion. Perhaps the universe is actually consistent for only relatively short periods of time, say millions or billions of years (long for a human, but short for a universe). Yet in this assumption of universal consistency lies the basis of all modern scientific inquiry. You can do scientific research in the laboratory and have every expectation that your results will apply in the world outside the lab. You can reverse the process with the same expectation. Similarly, you can do an experiment today and expect that your results will replicate those of a similar experiment of a century ago. Or, you can read a scientific account from a century ago and figure that the natural principles that governed the phenomenon back then are the same principles that are at work today. One cannot be a scientist without believing that the laws of nature are universal. Yes, individual events do vary from time to time and from place to place (otherwise, it would be a very dull world), but the *principles* that govern change are, in the Newtonian view, absolute and universal.

On this premise of universality Newton built an elegant mathematical structure that even three hundred years later continues to present considerable challenges to modern readers, particularly college students. The details can be found in numerous books,[1] including Newton's.[2] Using a very broad brushstroke, we can summarize Newton's most important scientific contributions as follows:

> 1. Many of the millions of events that occur daily are precisely and mathematically predictable.
> 2. When two observable systems interact, the interaction always goes both ways. One system cannot affect another without the second exerting an observable and predictable effect back on the first.

The resulting Newtonian Revolution yielded impressive gains in our understanding of nature. Advances were quickly made in the study of fluids, chemistry, and heat, and later, following the same traditions, in microbiology and electricity. The implication was clear that the universe was a highly orderly place, and that confusion was merely a human mental state. With enough diligence and patience, humans could eventually tease Mother Nature into giving up all her secrets, and, when this grand scheme of things was finally known, everything that happened afterward would be predictable. The universe was compared to a giant integrated clockwork, its gears meshing and pendula swinging in precisely regulated synchronicities. This Newtonian perspective gave promise that educated societies would soon be liberated from the capricious effects of natural forces run awry.

Then, on that disastrous November morning less than thirty years after Newton's death, the grand city of Lisbon was suddenly shaken into rubble, partly washed away by tsunamis, then ravaged by fire: an event completely incongruous with any other in living memory, a cataclysm that jumped out of nowhere, totally disconnected from every previously observed pattern in nature. This terrible disaster was unaccounted for by any of modern science's theories or laws, let alone predicted by them. Some forty thousand humans perished when, had they been able to see into the future by even an hour, they might have survived. Had they been able to see into the future by even a day, food supplies and ships could have been saved, and there would have been no fire. It was slim consolation that Newtonian mechanics could successfully predict the precise positions of the planets.

True, an earthquake had struck Lisbon a few centuries earlier (1531) and also killed thousands on that occasion. But that was before the Enlightenment, when humans were still ignorant of the laws of nature. Documentation of the earlier event was sparse, and eighteenth-century thinkers could assume that the sixteenth-century disaster must have been preceded by natural signs that had gone unheeded by the city's prescientific population. Now, the catastrophe of 1755 suggested that warnings are not always to be expected. Nature can indeed turn civilizations topsy-turvy without the courtesy of sending prior notice. It was not a time to be too smug about Newtonian successes in unraveling the mysteries of the universe.[3]

The universe has its clockworks, but it also has its tigers. Are Newtonian equations capable of predicting what happens if a tiger gets its tail caught in a moving train of precision gears? Will other equations, more refined, ever be capable of rendering such a prediction?

Ancient Forecasting

Let's backtrack a bit in time and examine how the very notion of pre-dictability may have originated. Etymologists tell us that the word "disaster" comes from the combination of *dis-* (unfavorable) and *astro* (stars). To the ancients, a disaster was literally an event precipitated by bad stars. This word origin is more than a piece of trivia; it begs us to inquire why the ancients chose to link stars, those unreachable remote points of light, with events that affect humans living on Earth.

The ruins at Stonehenge, England, are but one of many archaeological testimonies to prehistoric humans' belief in a connection between the heavens and Earth. This giant astronomical calendar, dating from around 1500 B.C., is a 91-meter (300–ft) circle of huge upright stones, many of which seem to have been moved tens of kilometers to the site. Clearly, to invest labor on the scale of Stonehenge, or the Mayan observatories in the Yucatan, or the less well-known Anasazi observatories in the southwestern United States, the leaders of these prehistoric societies had to have been quite convinced that the stars have something to say of human relevance.

In fact, the evidence was far from subtle. In Egypt, the flooding of the Nile, which renewed the nutrients in the farmlands and permitted the devel-opment of cities that didn't need to move after a few growing seasons, was signaled annually by the first appearance of the same constellations above the horizon. Year after year, century after century, when the stars moved into certain positions in the sky, the flooding soon followed. In fact, the very con-cept of a "year" probably grew out of this kind of observation, rather than preceding it. Meanwhile, at more northerly latitudes, the hunter-gatherer societies discovered that the migrations of the wild herds and flocks of birds were correlated with the reappearance of certain star patterns in the heav-ens. Many widely separated prehistoric societies would have found their survival to be synchronized with astronomical cycles. No doubt it was early humans' faith in the immutability of these annual cycles that gave them the confidence to first settle in regions with bitter winters, for they knew that even the worst winter would be of finite and predictable duration.

I am not, however, suggesting that correlations between astronomical and terrestrial cycles would have been particularly obvious. The star Arcturus rises at sunset, and a few days later the geese return. Nothing is written down. The two events happen again, roughly 365 days later, after thousands of other daily human observations and experiences. Would the average person immediately make the connection between the star rise and the geese? I think not.

But there were many thousands of years of prehistory, and hundreds of thousands of potential observers in any given year. Eventually someone was bound to notice such curious correlations. Inevitably, as the word spread, some curious person transcended mere observation and figured out how to predict when the geese or the floods would *next* return. He or she was humanity's first scientist. The essence of science is to predict.

Through such successful predictions, the concept of a "year" was born. If an early human patiently notched a stick each day, or put pebbles into a pile, he or she may have found that there were always 365 day–night cycles in one annual cycle. But it wasn't necessary to actually count the days; in fact, it was easier to use astronomical markers. If you watch the sunrise every day, standing at the same observation point, you find that the Sun does not always peek out above the same spot on the horizon. In the Northern Hemisphere during the spring, the Sun rises a bit farther to the north each day until it reaches a northernmost rising point. This day is the summer solstice, the day with the lengthiest period of daylight. In subsequent days, the sunrise shifts southward along the eastern horizon until a southernmost point is reached. This is the winter solstice, the day of the year with the shortest period of daylight. Individuals who first developed an understanding of these patterns undoubtedly became valued and supported by their societies for their ability to predict the future.

There was another, shorter, astronomical cycle that had a terrestrial correlation. In societies that derived their livelihood from the sea, the flow and ebb of the tides were of considerable importance. Ships left harbor on the outgoing tide and sailed in on the incoming tide. If a storm coincided with a high tide, shoreside communities prepared for flooding. From ancient times, the tides were observed to correlate with the position of the Moon: If you saw the Moon at a particular position above the horizon at high tide, you could be sure that another high tide would occur when the Moon again appeared in the same position, a little more than a day later. This much was straightforward enough. However, between these two events there would be *another* high tide, preceded and followed by its own low tides. Thus, the position of the Moon only told you when *alternate* high tides would occur. Superimposed on this pattern, every 15 days the tides are especially high (and, as they recede, particularly low). This latter cycle correlates with the observation that the Moon goes through a complete set of phases, from full Moon to the next full moon, in 29.5 days. Extreme tides are observed when the Moon is in its full and new phases (although not necessarily directly overhead). The evidence was clear that the heavens influence the seas in a manner that is at least partly predictable, though the pattern is by no means a simple one.

Up to this point, the fledgling science of astronomy was on firm footing. It was even quite reasonable, in the context of prehistoric and ancient understanding, to hypothesize that *all* astronomical cycles are correlated with cycles of terrestrial events. Hypotheses are fine; this is how science progresses.

What other astronomical cycles might be of interest? The only ones remaining take longer than a year: the appearance of the visible planets in particular constellations, alignments (syzygies) of the planets, solar eclipses, periodic appearances of comets, and so on. Although such events take place quite often on a cosmic calendar, they are relatively infrequent within a human lifetime. Thus, if a large comet appears in the sky during the months when an empire is being conquered (as happened in A.D. 1066), an observer might hypothesize that empires fall when comets appear. This is not pure speculation, because it has an observational basis. The problem is that the hypothesis needs additional validation, and the original observers are most likely dead before the comet reappears seventy-six years later, when no empire falls. Human nature being somewhat impatient, we jump to conclusions rather than wait.

Because the remaining astronomical cycles of interest were too long to permit anyone to validate or falsify any predictive hypotheses within a human lifetime, the fledgling science of astronomy inevitably gave birth to the mystical pseudoscience of astrology. Although astrological horoscopes may have a reassuring mystical value for some, they are clearly a dead-end street for those who would make sense of the dynamics of the universe, for they have long been divorced from their roots of predictive validation. To succeed in explaining disasters, we must go beyond any considerations of "unfavorable stars."

Numbers and Nature

Numbers themselves are abstractions; they are products of the human mind that don't exist in nature. Still, we often find it useful and meaningful to attach numbers to natural things. We can do this in two ways: through counting, or through measuring.

Counting is straightforward, a matter of placing our system of integers (whole numbers) in one-to-one correspondence with a set of observed objects. Thus we can count 14 ducks in a pond or 56 people in a room. We don't get fractional ducks or fractional people. Whether we use the metric system or the U.S. system, our answer is no different. Fourteen ducks metric are 14 ducks U.S., and that's that.

Yet Mother Nature does not seem to restrict her quantitative patterns to the integers. Around 510 B.C., on the Greek island of Samos, the philosopher Pythagoras did some novel experiments with stringed musical instruments.[4] Pythagoras took a multiple-stringed instrument and adjusted the tensions until all of the strings produced the same musical pitch. Then he placed a bridge under a string at one-half its length and found that when he strummed the string, the sound was in harmony with an unbridged string. Yet when he moved the bridge just a little off the halfway point, he found that the resulting sounds were discordant. Continuing the investigation, he found musical harmonies when the bridge was placed at one-third, one-fourth, one-fifth, and one-sixth the string length. Previously, only integers had been associated with natural objects, but now, for the first time, Pythagoras found a relationship between a physical phenomenon (sound) and fractions (an abstract mathematical creation of the human mind). It was one thing to cut a loaf of bread into thirds, imposing a pattern you had created mentally onto the external reality of the bread. It was a much more profound thing to have a stringed instrument tell you it prefers to be bridged in thirds, or halves, or quarters, because otherwise it sounds terrible.

But suppose that the bridge is positioned at a point that results in a sound that is totally unharmonious with an unbridged string. Isn't this point, wherever it is, also positioned at some fraction of the string length? Pythagoras struggled with this idea. Is it possible to have two string lengths whose ratio *cannot* be described by integer fractions? The answer, it turns out, is yes. There are many places (actually, an infinity of them) where one might bridge a string and not divide it according to a ratio of integers. Integers and fractions do not account for all of the numbers one might want to use in describing nature. Numbers like π and $\sqrt{2}$ are needed too, if for no other reason than to describe the bridge positions that give awful sounds.

Pythagoras and his students gave rise to the tradition that we should expect to find mathematical patterns in nature. Equally important, he showed that a complete description of such patterns cannot rely solely on the counting numbers (the integers). If we want to describe natural phenomena quantitatively, we occasionally need to invent new mathematical abstractions.

Pythagoras also made it clear that we can never hope to understand nature if we restrict ourselves to counting. An entirely different process is also needed: something we now call *measurement*. By measurement, we mean the comparison of a physical quantity with a standard. The numerical result of this comparison depends on the standard used, so that a person's height may simultaneously measure 67.5 inches, 171 centimeters, 5.63 feet, or 0.001065 miles. We choose the standard that best fits our purpose. Yet we must keep in mind that none of these representations is exact, for the probability of getting an exact numerical relationship when making a comparison is virtually zero. An "exact" measurement is always a contradiction in terms.

In the Pythagorean tradition, modern scientists spend much of their time struggling to find numerical patterns in nature. Scientists realize, however, that numbers themselves are quite artificial, falling far short of describing any fundamental truths of the universe. What is important to science is how the relevant numbers are interpreted, and what these interpretations imply about the course of future events.

A Giant Step Backward

The traditions of modern scientific thought are often miscredited to the Greek philosopher Aristotle (384–322 B.C.), a student of Plato and tutor to Alexander the Great. Normally one does not hold the professor responsible for the misdirected ambitions of his students, and in this spirit we can excuse Aristotle for having taught one of history's most megalomanic conquerors. Alexander's empire quickly crumbled after his early death, and in the long run his life was largely irrelevant. Aristotle had a more lasting impact: Some of his writings continued to retard progress in science for the next eighteen hundred years.

Aristotle's surviving writings describe science as a purely intellectual process, whose objective is to identify the absolute truths of the universe. This process begins with the observation of "particulars," the individual events that happen in our life's experiences. From here, Aristotle develops "universals," or abstract truths common to a class of particulars. When enough universals have been identified, they are combined to get higher-order universals. Ultimately, this process should lead to the "first principles," or the truths that cannot be explained in terms of anything else. Few modern scientists would disagree with this part of Aristotle's conceptual scheme.

Aristotle tells us that before we can climb his intellectual pyramid, we first need to classify things according to his "categories." We then apply Aristotle's rules of logic, which become somewhat complicated in their detail but which are fundamentally based on the axiom that "a thing cannot both be and not be in the same manner at the same time." A person is either bald or not bald, for example. (No "fuzzy logic" allowed here, no fellows that might be balding.) On a more profound level, the paradigm of gradual change is wiped out *a priori*. Natural evolution would be an impossible conclusion of Aristotelian science.

The flaws, however, are much more serious than this. Aristotle's observations of "particulars" were quite informal; he did not advocate experiment or measurement in the tradition of Pythagoras. Aristotelian science had nothing to do with manipulations; it had only to do with the mind. He wrote, for instance, that heavier objects fall faster than lighter objects, when a very simple experiment could have confirmed that weight is not the critical variable in the rate of an object's free fall. In fact, Aristotle's own logical system could have told him this; the *reductio ad absurdum* logic might have run as follows:

Suppose that a 10-pound object falls faster than a 5-pound object.
But a 10-pound object is the same as two 5-pound objects stuck together.
Therefore, a 10-pound object falls faster than either of its two halves.
But this is an internal contradiction.
Therefore, the initial premise must be rejected.

No, Aristotle himself never advanced this argument, and that is my point. He restricted his proofs to premises he already thought to be true, and, apart from some relatively trivial biological observations, his volumes of writing failed to produce a single new scientific insight of any importance. His failure to consider the value of controlled experimentation reflected his own social status: the elite of Athens of his day simply did not perform manual tasks. Beyond this, to have had to experiment would have contradicted Aristotle's world-view that the life of the mind alone leads to the highest truths.

In his rejection of validative experimentation lay the most serious problem with Aristotle's scientific methodology. For, once one arrives at universal truths, what test is there? Aristotle's truths, once attained, were absolute, and no test was needed. Anyone who questioned the observational basis of such truths was merely displaying inferior intellectual ability and was not to be taken seriously. When Aristotle decided that the earth was the center of the universe, to him and his many followers this truth was absolute, and the matter was settled forever. In practice, however, such matters were settled for only the eighteen centuries it took for a new way of doing science to come along.

Science and Authority

Around the year A.D. 1250, copies of Aristotle's writings fell into the hands of Thomas Aquinas, a Dominican friar. The idea of proving the truths of the universe through pure logic, in a manner that was forever immune from further challenge, had tremendous appeal. Aquinas used Aristotelian logic to prove all the teachings of the Roman Catholic Church, and, with these exhausted, went on to prove a vast compendium of new theological truths that no one had ever thought of before. By the time he died, at age 49, Aquinas's *Summa Theologica* had run to several dozen thick volumes. The papacy endorsed the work in its entirety and, along with it, all of the Aristotelian writings the *Summa* was based on. What had begun as a quest for scientific truth was now dogma, to be believed under pain of excommunication, or worse.

Yet even Aquinas and the Church authorities had not anticipated every possible human thought, and there were enough gaps in Church teachings to allow some new discoveries, particularly in physiology, alchemy, and the more applied fields of mechanics and structural design. Around 1340, in England, William of Occam proposed a criterion that would help one to choose between alternative explanations for the same phenomenon. This criterion, which for murky reasons is still referred to as "Occam's Razor," may be stated as follows:

> When several conflicting explanations are proposed for the same set of observations, the best explanation is the one with the fewest independent assumptions.

Note that Occam's Razor does not refer to "truth," but only to the "best" explanation. This was a giant step forward.

As a simple example, let's return once again to the Lisbon disaster. One hypothesis is that there were three separate events: (1) a monumental storm somewhere in the Atlantic that sent great waves crashing toward the Portuguese coast, while at about the same time (2) an earthquake hit Lisbon, while nearly simultaneously (3) a fire broke out in the city. A second hypothesis suggests that there was a single large earthquake beneath the sea floor off the coast of Portugal, and that this event was responsible for all three of these effects. This second hypothesis, by linking the seismic waves, tsunami, and fire to a single origin, is surely the better, or more probable, explanation. But is the second hypothesis true? There is no way to tell for certain. All we can do is settle for "best," on the basis of the evidence we happen to have available.

Modern science does not even pretend to uncover absolute truths. Instead, it generates hypothetical explanations, then sifts through them so that only the simplest survive. By "simplest" we don't necessarily mean the easiest to comprehend. Rather, we mean simple in the sense of relying on the fewest independent assumptions. Einstein's theory of special relativity, for instance, explains an enormous variety of events on scales ranging from the subatomic to the cosmic, by beginning with but two basic assumptions. In this sense it is marvelously simple, for no competing theory can explain so much in terms of so little. The theory of relativity, however, is hardly an intellectual breeze.

Mikolaj Kopernik may not have heard of Occam, but in the early 1500s this Polish cleric used the same razor to slaughter one of the sacred cows of the Aristotelian tradition: the earth-centered universe. By A.D. 200 a great infrastructure of mathematics had been built to explain the observed motions of the planets from a presumed stationary Earth, and in the refinements of the following centuries the workings of this mathematical model came to depend on dozens upon dozens of independent assumptions. Kopernik (who latinized his name to "Copernicus" in his writings) showed how the same planetary observations could be explained quite easily by treating Earth as just one of the planets, all of which circled the Sun. In Occam's tradition, Copernicus was careful to say that this was not necessarily the "truth," but rather a much, much easier way of looking at things. Exercising additional caution against upsetting Church authorities, Copernicus delayed publication of his ideas until 1543, when he lay on his deathbed.

Copernicus studied in Italy and lived in eastern Europe, where the authority of the Catholic Church was immutable. But in northern Europe and Britain lived societies that had broken away from papal authority. Work-

ing in Germany and Denmark, Johannes Kepler (1571–1630) was under no authoritarian religious pressure to continue to believe Aristotle.

Kepler believed that all past events, and all that would occur in the future, had been programmed into the matter of the universe at its moment of creation. Therefore, if your understanding is valid, you should be able to predict the unfolding of very specific future events. If such a prediction is borne out by the actual course of events, your understanding is confirmed. But if the predicted event does *not* occur, you must be willing to reject your explanation, formulate a new one, and repeat the process. Using this criterion over a period of several decades, Kepler finally succeeded in formulating three rules (or physical laws) that permitted anyone to predict the configurations of the planets at any time in the future, to an accuracy limited only by visual acuity. Interestingly, Kepler also supported himself financially by casting horoscopes for members of the nobility, an endeavor in which his predictions were not nearly as accurate. Until his death, though, he believed that all events, including human futures, are in principle quite predictable.

Meanwhile, in Italy, Galileo Galilei (1564–1642) was adding another dimension to the process of scientific inquiry: that of controlled experimentation. For the previous eighteen centuries teachers and professors had been telling students, on the authority of Aristotle, that heavy objects fall faster than light objects. Yet over all these years no one seems to have tested the theory with an experiment (or else dared to speak of the findings). Galileo climbed to the top of the Leaning Tower of Pisa, leaned over the railing, and simultaneously dropped pairs of balls of different weights. In doing this, he was careful to ensure that the objects differed *only* in their weights. He did not compare cannonballs and feathers, for instance, because such objects differ in several obvious characteristics other than their weights. Galileo recognized that if the variable of interest is weight, then that is the only variable that should be varied. His famous result: All other factors being equal, heavy objects fall at the same rate as light objects.

Galileo went farther with this idea of controlled experimentation and studied a variety of other motions, including that of a weight swinging at the end of a string. His finding that the period of a pendulum's motion depends only on its length and not its weight quickly led to technical spin-offs in the design of precision clocks. Then Galileo moved into astronomy and got into serious trouble with Church authorities. After improving the telescope, which had recently been invented in Holland, he turned it to the heavens and discovered the phases of the planet Venus and four of the satellites of

Jupiter. This meant that there were heavenly objects that were *not* orbiting Earth – a finding clearly incompatible with the theory of a geocentric universe. Galileo proudly proclaimed, to everyone who expressed the slightest interest, that once and for all he had disproven the theory that Earth was the center of everything. This, however, resulted in a stern warning from papal officials, who saw their authority being undermined. In response, Galileo wrote his *Dialogue on the Two Chief World Systems*, in which Simplicius, a thinly veiled caricature of Pope Urban VIII, ineffectively (and stupidly) attempts to defend the geocentric theory. For this Galileo was brought to trial, forced to recant his heretical teachings, and kept under house arrest for nine years, until his death in 1642 at the age of 78. Only in 1992 did the Church formally pardon him. Galileo's punishment, however, was light, compared to that of his fellow Italian scientist, Giordano Bruno, who was burned at the stake in 1600 for a similar scientific heresy.

To the credit of the Church, it did learn a lesson from these dark episodes, and Church authorities implicitly acknowledged this long before they pardoned Galileo in 1992. When, in the mid–ninteenth century, science again pulled the rug from under a major tenet of Christian dogma, that of the historical truth of the creation story in Genesis, the Roman Catholic authorities declined to take an official position. Catholics were left free to accept or reject the theory of evolution on their own intellectual terms. Meanwhile, unfortunately, the religious authorities of several more recently evolved Christian denominations did all they could to suppress the teaching of theories of biological evolution. This is more than a little ironic, given that Protestantism owes its very origins to individuals who valued their own thinking more highly than the views of the prevailing Christian religious authorities. Maybe some day the "creationists" too will learn this lesson of history: Mother Nature doesn't worry about conforming to mere human expectations. She does what she does, and if we want to understand, it is up to us to pay attention to *her*, not to mere human authorities or their institutionalized versions of absolute truths.

Cause and Effect

One legacy of Aristotle that is still largely with us is the notion of cause and effect. We witness an event, and we presume it happened *be-cause* of something else.[5] For Aristotle, the cause and the effect had to be present at the same point in space, with the cause preceding the effect by only a very short time. These *ad hoc* requirements of simultaneity and locality were necessary to prevent one's saying, for instance, that day "causes" night and night "causes" day.

The notion of causality, however, does not appear in Galileo's works, nor is it a major theme in Newton's writings. To Newton, there is certainly a *link* between ocean tides and the Moon, but the link goes both ways. One object does not influence a second unless the second also influences the first. Which is the cause, and which is the effect? It's the observer's call. If you're interested in the tides, you might say that the Moon *causes* the tidal cycles. But if you're interested in the Moon, you will say that Earth's tides *cause* the Moon to gradually change its orbit. Cause–effect labels are simply a matter of one's perspective. Nor does Newtonian mechanics require local proximity between effects and their causes; after all, the Sun and Earth manage to interact gravitationally across 93 million miles of mostly empty space. An event in one place may be linked quite strongly to a second event a great distance away or, for that matter, quite some time later.

Yet the notion of cause and effect continues to survive, both for practical convenience and "because" these concepts are intricately woven into the fabric of our language. The events forming a Newtonian event pair are seldom of equal importance; if a seismic wave comes along and a city collapses, our natural human perspective is that the human disaster was the "effect" and the seismic wave was the "cause." We don't care a great deal about what the falling city did reciprocally to diminish the wave. This quite natural anthropocentric viewpoint can, however, lead to serious oversights. A swollen river devastates a growing city, and we view the flood as the cause of the devastation. We are not inclined to question whether the presence of the city itself may have caused the flood (through local deforestation, for example, or through construction of dikes that prevented the river from dispersing onto a natural floodplain).

Determinism is the view that all events arise unambiguously from well-defined causes: that specific natural agents acting on specific natural systems have no option but to produce predictable effects. From a deterministic perspective, if an unanticipated event occurs, it is because either (1) we didn't sufficiently understand the natural chains of cause and effect, or (2) we didn't pay enough attention to our observations. In this view, Mother Nature doesn't flip dice; she precisely programs all her actions at all times and in all places. If we can unravel the mind of Mother Nature, then, we should be able to follow her program ourselves and successfully predict every event that will ever occur, in all places and for all time. In the deterministic view, the future was programmed – completely and irrevocably – at the moment of the creation of the Universe. Do I believe this myself? Not quite.

But the goal of eighteenth- and nineteenth-century science was no less than this: to eliminate any chance that a natural event would take someone by surprise. Scientists successfully described vast segments of Mother Nature's master program – laws explaining electricity and magnetism, waves, heat, light, sound, and the motions and interactions of atoms. By the 1890s, virtually every practicing scientist was a determinist in his or her world-view. The fledgling science of psychology embraced a similar premise that every human action and every human thought were programmed by the interactions between the individual and his or her environment. All reality, from the atoms to the stars, moved and interacted in a cosmic clockwork. Humans, composed fundamentally of atoms, were no exception. Note that there is little room for free will in this world-view.

In fact, by 1890 one *could* predict the future, and confirm such predictions, in a vast variety of controlled laboratory experiments. Although such predictions did not work quite as well in the world outside the laboratory, most worked well enough to permit the rise of an engineering profession that could justifiably busy itself with design calculations. No longer did an engineer build a bridge and then stand around watching nervously to see if it would collapse (as did up to 25% of the bridges built in the 1870s).[6] Instead, it became possible to *predict* the performance of a bridge before its plans ever left the drafting table. Tremendous investments of capital could be solicited on the strength of predictions that, for instance, a city could be lighted by tapping the energy of a local waterfall. Scientific predictability became the foundation of modern engineering.

So if science could go just a bit farther, make a few more links between the natural laws it had already discovered, account for just a few more of the intransigent variables of nature, then nature would never again have the ability to serve up any kind of unpleasant surprise. Disasters would be averted by technologically based prediction and response. We would always know what was about to happen and would use this to human advantage. (Never mind the obvious contradiction between determinism and decision making; this would be left to the philosophers and psychologists to argue.) Such was the situation in the 1890s: The scientists committed to a deterministic view of the cosmic order, certain that the last few missing links were very close to being discovered.

Then in 1896 came Antoine Becquerel's discovery of radioactivity, a distinctly nondeterministic phenomenon, followed in the next few decades by the discoveries of a whole variety of other fundamental processes that were equally unpredictable in their outcomes. One could talk statistically about the average lifetimes of the energy states of atoms, or half-lives of radioisotopes, but one simply could not predict when (or in some cases where) a particular subatomic event would occur. Some felt that this implied the existence of other hidden variables that the scientists had so far missed, and even the great Albert Einstein proclaimed at a 1927 scientific conference in Brussels, "I am convinced He does not play dice," the "He" being Einstein's metaphor for the creator of the universe.[7]

Meanwhile, others did experiments whose outcomes clearly suggested that nature, at its most fundamental level, is inherently nondeterministic. On human scales of time and space, nature may sometimes *appear* to be deterministic, simply because of the law of averages as it applies to large numbers of individual particle interactions. As of this writing, we now have a century's worth of evidence, in uncountable thousands of experiments, to suggest that searching for new variables does not erase the inherent indeterminacies of subatomic particle interactions. In fact, all of this experimentation has led to the discovery of whole new classes of phenomena that fail to yield to deterministic prediction.

What relevance might nondeterministic processes have to the theme of natural disasters? Let me ask you to take a big leap of the imagination for a minute or two and mentally expand the nucleus of a radium atom to the size of Earth. Such an object might make a perfectly good planet (although its gravity would certainly be a bit strong). On this planet, put a life form – say a genetically engineered very strong species of ant. My question: How long would this ant colony survive before it was destroyed by a cataclysm?

Before you object that this is a dumb question, let me beg you to stay with me a moment, and let me point out that I can describe the nature of the cataclysm: Without any warning some evening, a chunk amounting to a bit less than 2% of the planet's mass suddenly bursts from its surface and shoots off into space. The planet recoils with an incredible jolt, and its core redistributes itself to fill in the hole, releasing great additional quantities of energy. When things finally settle down, the new planet looks quite a bit different, and the ant life on its surface has been destroyed. Measurements made on samples of large numbers of radium–226 atoms tell us that half of them will undergo the process I just described in about sixteen hundred years. (The chunk of material ejected is called an "alpha particle".) But I am asking you to be more specific. How long will *this particular* ant-inhabited radium nucleus last?

You'd like more information? I describe the force (called the "strong nuclear force") that holds the pieces of the nucleus together. I describe the repulsive force (called the "electrostatic force") that is responsible for ejecting the alpha particle. We dig deeper, breaking apart some of the protons and neutrons in the planet and finding that they're composed of still smaller particles. But none of this helps. The best we can say is this: If we had one hundred of these radium planets, about fifty would still be around in sixteen hundred years, whereas the other fifty would have experienced a major catastrophe.

If all we have is one radium nucleus, it may still be around in 1 million years, or it may blow its top tomorrow. There is no way to predict, with just one. All we can predict is statistical probabilities, which represent nature's general trends with large numbers of events over large time scales. It's important to note that probabilities are not being used here as a refuge for those who are ignorant of the finer details; rather, the view is that natural processes themselves are inherently statistical at their most fundamental levels. Einstein objected to this idea. Most of today's scientists accept it. There is today compelling accumulated evidence that nature's determinism is at best statistical.

But can the physics of the atom really have anything to do with large-scale phenomena like hurricanes, earthquakes, or epidemics? It can and it does, in this sense: All physical and biological events are composites of uncountable billions of submicroscopic interactions. Because the outcome of each of these underlying interactions always has a built-in degree of indeterminacy, it follows that all large-scale composite phenomena will also have fuzzy outcomes. How fuzzy? The question can be answered for a variety of specific controlled experiments in the laboratory, but for the general case of large-scale natural phenomena, we are still woefully ignorant.

As we dig deeper into the mysteries of nature, we find that the Cosmos, on all scales of size, is much less deterministic than scientists once believed. Earlier deterministic approaches were successful only because they keyed into those events whose outcomes occur with nearly 100% probability – the motions of planets, for instance. Newton would never have made his important scientific contributions if he had begun by studying earthquakes, windstorms, or the epidemic that closed his university in 1666.

The philosophical position that individual events may be in principle unpredictable, while averages and other statistical indices of large numbers of similar events are predictable, is referred to as *statistical determinism*. The important point here is that predictions are still possible, provided that we are willing to settle for statistical predictions. As for "cause and effect," one must drop the notion completely before passing through the gate to statistical analysis.

The Fragmentation of Modern Science

As we have seen, science grew out of the social imperatives of prehistoric and ancient civilizations to predict, and plan for, the future. Most early societies that gave birth to science seem, independently, to have eventually institutionalized it in a priesthood. Such institutionalization, by replacing protoscientific inquiry with officially sanctioned dogma, virtually guaranteed stagnation. When a point is reached where one thinks he or she has all the answers, there can be no science.

In ancient Greece, however, science was allowed to fall within the scope of the intellectual activities of the philosophers. Some, such as Pythagoras (and later, Eratosthenes and Archimedes) considered it quite natural to make controlled measurements and to perform experiments to test their conjectures. Others, such as Aristotle, considered reason alone to be the appropriate investigative tool. But for all these men (and yes, except for Hypatia all were men), any question about anything was fair game for science. Science was not a mere collection of facts or findings; it was a process of *seeking*.

Aristotle's writings, as we have seen, became institutionalized by a different kind of priesthood and the inevitable stagnation followed. Only with the Enlightenment of the 1700s did a social climate develop that nurtured widespread interest in scientific inquiry.

Thinkers noticed that some categories of scientific questions yielded to an empirical (that is, experimental) approach, using formal methods of hypothesis testing. Such questions, which are today discussed in books on astronomy, physics, and chemistry, were said to constitute "natural philosophy." Other kinds of questions were better examined through detailed field observations of natural environments and by analyzing the written accounts of explorers of far-off places. These questions fell under "natural history," which included what we today would characterize as biology, geology, geography, and their subfields.

By the middle of the nineteenth century, the successes of natural philosophy and natural history had resulted in so many observations, hypotheses, laws, and theories that no single human mind could possibly master them all in a lifetime and still have time to think about eating. The fields of natural philosophy and natural history were inevitably fragmented into disciplines, and then into subdisciplines. Today the scientific job ads solicit such specialists as a "theoretical condensed matter physicist" or a "physical oceanographer, nonlinear acoustics." No longer does an employer figure on hiring a generic scientist. With so many field fragmentations, what could someone who still called him- or herself simply a *scientist* in the 1990s possibly know?

A consequence of the success of modern science is that geometrical increases in human knowledge have far outpaced the modest increases in the life span of the typical scientist. We assimilate new ideas arithmetically in time; that is, if it takes a week to master a new idea, it will (on average) take two weeks for two new ideas and three weeks for three. Meanwhile, hundreds of new ideas appear in print each day, and the number keeps growing. Each day we spend learning, we know less of what there is to know. The only way to make any progress is to specialize. But when everyone specializes, the sciences fragment into a multitude of subdisciplines, each with its own dialect. This presents considerable challenges for one who wishes to examine a topic that cuts across the boundaries of the established scientific disciplines.

Fragmentation of the sciences has had additional unfortunate consequence: a widely held misconception that science is about "facts" rather than process. When someone uses the word "geology," for instance, most of us get a mental image of rocks. Maybe we think of the rows of mineral specimens we've seen in museums or the geological formations we've noticed in our own travels. What does *not* immediately spring to mind is the laboratory technique a geologist might use to inquire about the age of a sample of pumice, or the analytical technique she might use to question Mother Nature about a possible hidden geological fault line deep within the earth. The lay public, the court system, and the disaster planners all turn to scientists for answers, and if the "right" answers aren't provided, the scientists are often viewed as having failed. To the scientist, however, the *lack* of answers is the *raison d'être* for further scientific inquiry. It is questions, not answers, that excite and mobilize the scientific mind.

Yet modern scientists must also cope with their own cultural fragmentation. Let's briefly examine the major scientific disciplines and see how the boundaries first arose.[8]

> *Astronomy* was the earliest and simplest true science. It continues to be the most general science, in that it encompasses the whole Universe. Originally it dealt with only two variables: position and time, and a few

thousand observable objects in the skies. Although early astronomy succeeded in making extraordinarily accurate predictions about most extraterrestrial events, it left a gap of ignorance regarding events happening close to or on Earth.

Physics may be viewed as the most basic science, in that it restricts its attention to events that can be described through a combination of only seven variables (position, time, mass, electric current, temperature, luminous intensity, and a unit of atomic substance). The number of observable objects, however, is effectively infinite. Physics succeeds very well when the number of interacting objects is three or fewer. Put four particles together, though, and the predictions start to get fuzzy. Quantum physics adequately accounts for the existence of the hydrogen and helium atoms, but it fails to predict the precise structure and chemical properties of lithium (element no. 3) and all of the higher elements.

Chemistry skips over the problem of predicting the existence of the particular kinds of atoms and starts off by accepting the existence of 107 elements and their isotopes. Drawing upon those laws of physics that seem to help, chemistry develops theories of how atoms combine into complex molecular configurations. Although chemistry can predict the outcomes of a wide range of reactions, including some of those that drive the life processes, this scientific discipline fails to predict life itself. The nature of life remains a gap of ignorance between the fields of chemistry and biology.

Biology assumes life as a given. Life does not appear spontaneously; it arises only from old life, and there is a lot of old life on Planet Earth. Biology therefore deals with a very large number of variables, and most of its theories lead to relatively fuzzy statistical predictions. Fuzzy though they are, such predictions are enormously important in fields ranging from medicine to agriculture. No biological theory, however, comes close to explaining the conscious mind.

Psychology assumes that humans (and probably most higher animals) have a continuity of mind and memory, regardless of the fact that the organism's constituent atoms and molecules are continually being exchanged with the environment. From this starting point, psychologists seek to explain how we think, how we learn, how we love. The number of possible variables is so large as to defy classification, and the predictive theories are necessarily quite limited in their generality. The only quantitative analysis possible is statistical. Psychological theory provides a meaningful foundation for a wide variety of human activities ranging from education to baseball to disaster planning. The theories of psychology, however, do not explain how social institutions arise and evolve.

Sociology deals with interactions between groups rather than individuals, and with broad social and cultural patterns that take on a life of their own beyond that of the individual members of a group. Sociological theory seeks to predict broad social trends rather than individual behavior. Such predictions may be based on statistics, or they may begin with qualitative observations and end with qualitative conclusions. Because the number of possible observables is incredibly large, predictions based on sociological theory are quite tenuous and difficult to disprove. Nevertheless, the findings of sociologists are eminently relevant to those working in fields such as public policy and business.

We see from the above that the divisions between the scientific disciplines coincide with our greatest gaps of ignorance: Physics cannot predict chemical properties, chemistry cannot predict life, biology cannot predict consciousness, and psychology cannot predict the evolution of social institutions. Further, we should note (Fig. 2.1) that as we apply scientific inquiry to systems of increasing complexity (e.g., social systems), we sacrifice generality and predictive fidelity. Conversely, as we move to the sciences that deal with fewer variables, we can make predictions with better accuracy, and the related scientific theories can deal with more general classes of phenomena.

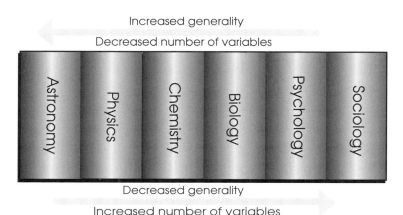

Figure 2.1. Fragmentation of the sciences. The divisions correspond to our greatest gaps of ignorance.

Obviously, I have left a lot out of this simple scheme. Some of the modern investigations in astronomy properly belong between chemistry and biology on the complexity scale rather than at the left of the band. Geology, which I haven't listed, draws primarily on the paradigms of physics but investigates systems that are in many ways more complex than those considered by chemists. The same is also true of meteorology. These examples nevertheless still support my point: that scientific inquiry into increasingly complex systems results in theories that have diminishing generality and predictive validity. With just two equations, one can predict the position of virtually any object in the solar system at any day and hour of interest for centuries into the future. Yet, with even a hundred equations, we cannot predict the rate of slippage of the San Andreas Fault in California next Saturday, or the noon-time temperature in Topeka four days from now. The most difficult challenges of science lie in attempting to understand complex systems.

Where does one place history in this scheme? It appears that much, and perhaps most, historical investigation is better classified under the humanities than under the social sciences. A scholarly description of a disaster, or even a documentation of a whole period of a country's past, is not science unless it advances a hypothesis that can be tested and ultimately generalized. Most historical investigation sticks to the subject and does not prognosticate. This makes the endeavor no less noble than a scientific inquiry, just different. And, as I have argued earlier, any scientific study of sporadic irreproducible natural phenomena can proceed only with great debt to the historians.

Truth in the Sciences

Most of today's scientists agree that truths are never knowable in an absolute or unconditional sense. Scientific inquiry can prove that someone's theory is *false,* but it can never prove that any theory, or even a simple observational premise, is unconditionally true. The only true premises science is capable of developing are those that are tentative and/or restricted in space or time.

Think of a scientific truth, even the most trivial, and try to state it unconditionally. The sky is blue? No, at least half the time it's black, and sometimes parts of it are red or orange. What goes up must come down? No, we've had space probes that have never returned, and helium is continuously taking a one-way trip from our planet into space. Penicillin kills bacteria? It kills some, but not all. Proceed from such simple statements to the development of a complete theory, and you are even more vulnerable to being contradicted.

In the early twentieth century, scientific thinking was dominated by the philosophy of *logical positivism*, which asserted that truth was indeed attainable through repeated and replicated experimentation and observation. By the late 1930s this view was seriously challenged, principally by the Austrian-born scientific philosopher Karl Popper.[9] Popper disputed that scientific inquiry could be reduced to a formal logical method (as the positivists suggested). Creative intuition is always an essential element of scientific theory building, and this places all scientific theories on subjective foundations, where they are intrinsically vulnerable. No theorist can ever afford to be so smug as to claim that he's bullet-proofed his theory against all possible future experiments that might disprove him. An experimenter can always develop empirical tests that will transcend the experiential base of any particular human who has proposed a scientific theory.

Our resulting view of the process of science is shown in Figure 2.2. At the base of the pyramid are the thousands of individual events we observe and experience through our human interactions with nature. From this hodgepodge of events, we seek to recognize recurring patterns that permit us to generalize about certain classes of phenomena (e.g., a falling barometer usu-

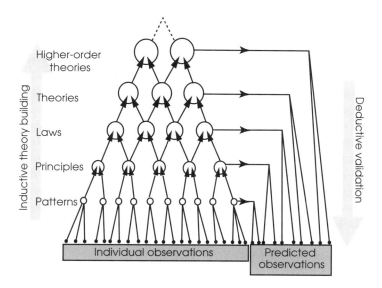

Figure 2.2. The pyramid of scientific inquiry. Theory building is an *inductive* process (recognizing the general from the specific), while validation is a *deductive* process (predicting the specific from the general).

ally signals an oncoming storm). We then try to group similar patterns together to develop a more general scientific principle. (E.g., liquids and gases tend to move from high-pressure areas toward low-pressure areas, and that is *why* a falling barometer signals an oncoming storm.) Numerous other principles can be formulated in like manner to describe other characteristics of liquids and gases, and these in turn can be combined into a more comprehensive theory of fluids, moving us up another level in the pyramid. This theory, in turn, can be combined with theories of other phenomena to develop a higher-order theory (e.g., how the atmosphere and volcanoes interact), and so on. The assumption is that somewhere at the top there will someday be a single theory of everything.

The process of science, then, begins with the particular and moves to the more general. There is no clear set of steps one follows to accomplish this; the process requires an intangible creative insight. As an admittedly artificial (but purposely simple) example, consider for a moment the following puzzle:

> An archaeologist unearths a wall in an abandoned village where there is evidence that the inhabitants spoke English. On this wall the following sequence of letters has been chiseled: O T T F F S S E N. Originally the sequence probably went on, but the rest of the wall is now gone. Is there a pattern here? If so, what were the next three letters?

Yes, I have an answer: T E T. But how do I prove I'm right? If you say N E S (under the assumption that the rest of the pattern reverses), how do you prove you're right? Clearly, there is no way to affirm the absolute truth of either of our conjectures simply by shouting our answers back and forth at each other.

So I go farther, and I explain how I arrived at T E T. If I count from one through nine, and I write down the first letter of the name of each integer, I get O (one) T (two) T (three) F (four) and so on. In this scheme, the tenth, eleventh, and twelfth letters are clearly T (ten) E (eleven) and T (twelve). But how did I arrive at this? What "scientific method" did I use? None whatsoever. I began by looking at the pattern over and over in bewilderment; I went on to something else; then, a few hours later when I wasn't consciously thinking about it, something clicked, and it occurred to me to try the first letters of the integers. And this is Karl Popper's point. The process of theory building is inductive, but it is not formally logical. There is no way I can lay out the steps of my solution so that you can critically analyze each step. A creative leap of faith was essential to my solution.

Still, I have not proven I'm right; your N E S could still be the answer.

What we both need to do is to find evidence of the next letter. We roll up our sleeves and begin digging in the ground at the end of the wall.

We luckily find the next stone, it mates perfectly with the intact section of wall, and it has a T, just as I predicted. Have you been proved wrong? You betcha. Have I been proved right? No, all we've done is to strengthen my theory. If we dig some more, and the next mating stone has a Q, my whole wonderful theory goes down the drain. The pattern of letters can still be something else (perhaps more profound), and there still remains the possibility that it is not a pattern at all but just gibberish. In fact, no amount of digging can ever prove my theory to be absolutely true; it can, however, certainly prove me wrong.

This is one of Karl Popper's main points. One never finally proves a scientific theory; all one can do for certain is disprove it. In fact, if you propose an idea but do not suggest how I might disprove it, then your statement cannot qualify as a scientific theory. It might be a philosophy, a religion, or some other world-view, and it might even have value and validity, but without a means of falsification it is not science, and no scientist will take it seriously. *Every scientific theory must have embodied within it the means by which it can be disproven.*

Now this may sound like a good program for getting nowhere: creating theories with great caution but destroying them with gusto. What it does, however, is to guarantee that any half-baked theory is quickly eliminated before too many other scientists build further upon it. Without the merciless practice of attempting to disprove each newly proposed theory, science would never have progressed beyond the intellectual shallows of Aristotle's writings. The essential ingredient that Aristotle did not provide was the deductive, predictive requirement. Looking once again at Figure 2.2, we see that at each level of the pyramid we must use our proposed theory (or principle) to predict the outcomes of events that have not yet been observed (and that therefore could not possibly have been part of the set of events that formed the original foundation of the theory). We then demand that the theoretical predictions be consistent with our observations as these new events unfold. If a theory fails this essential test of empirical verification, it must immediately be rejected (or at least modified). If, on the other hand, a theory's predictions do work out in reality, we let the theory continue to stand until the next empirical test.

A necessary consequence of this scheme is that no theory can ever be complete, and therefore no theory can be absolutely true, because *we cannot possibly observe the whole universe for all time.* There will always be events and phenomena we don't know about, and which our theory doesn't address. Some readers may view this situation with pessimism and disappointment, but to a scientist it is a wonderful thing. It suggests that there will always be science to do, far into the future.

To those who deal with natural disasters, however, this presents a serious dilemma. If scientifically generated "truth" is conditional, relative, and temporal, how can we possibly rely upon science when we write building codes for earthquake zones, develop evacuation plans for regions subject to hurricanes, respond to an outbreak of an epidemic, or establish a rate structure for disaster insurance? Won't all of our plans and policies, no matter how carefully considered, be flawed? The answer, of course, is that no public policy can possibly anticipate what scientists might learn in the future. When a major earthquake struck Kobe, Japan, on January 17, 1995, it destroyed nearly fifty thousand buildings and claimed more than five thousand lives.[10] Yet almost all of the structural failures were in buildings constructed before 1980, when authorities enacted a more stringent construction code, which, in turn, was based on some relatively minor scientific advances of the previous decade. Those who developed the earlier building code can hardly be blamed for the terrible losses, nor can the scientists of the 1950s, who advised the engineers and public policymakers of that earlier time. Time passes, and if we pay attention, we often learn. The 1995 earthquake itself provided scientists with an event for assessing the validity of current theories that link structural dynamics to seismic activity.

Data from Kobe will continue to be analyzed for many years, but, as a result, we will be even smarter in decades to come. And someday this new knowledge, albeit still imperfect, will filter into the arena of public policymaking, where it will result in improved codes and disaster plans that will save the lives of many who have not yet arrived on our planet. In the next chapter, we examine in more detail the interdependence of disaster science and structural engineering.

Notes

1   Some representative books that recast Newton's original formulations in more modern language and symbolism are the following: G. R. Fowles, *Analytical mechanics*, 4th ed. (Philadelphia: Saunders, 1986); A. Pytel & J. Kiusalaas, *Engineering mechanics: Statics and dynamics* (Glenview, IL: HarperCollins, 1994).

2 Isaac Newton. *Philosophiae naturalis principia mathematica* (Mathematical principles of natural philosophy) (1687; reprint, Berkeley and Los Angeles: University of California Press, 1962).

3 The French philosopher Voltaire was particularly troubled by this issue; see my comments in Chapter 1.

4 A natural outgrowth of the hunting bow (which gives a characteristic "twang" when shooting an arrow), stringed instruments arose independently in many different cultures. Pythagoras's experiment is nicely reconstructed in J. Bronowski, *The ascent of man* (Boston: Little, Brown, 1976), 155–7.

5 Aristotle identified four types of causes, of which only his "efficient" cause need concern us here. For Aristotle's original system of logic, as well as his other writings, see *The complete works of Aristotle: The revised Oxford translation*, ed. J. Barnes 2 vols. (Princeton: Princeton University Press, 1984).

6 The claim that 25% of the bridges built in the United States in the 1870s collapsed has been repeated in several sources and disputed in several sources. Having nothing to add to this argument, I will avoid it here. I mention this claim to point out that society's current expectations that engineers be infallible is a fairly recent development, and that through their historical improvement in their success rate engineers have been rewarded with increasing social pressures on their performance. Today's engineers cannot afford to be ignorant of the latest scientific developments in their fields.

7 D. J. Kevles, *The physicists: The history of a scientific community in modern America* (New York: Knopf/Random House, 1977).

8 The basic scheme I describe here was first proposed by the French philosopher Auguste Comte in 1842 in his book *Cours de philosophie positive.* I take some liberties that reflect both the developments of the last one hundred fifty years and my personal bias that mathematics (included by Comte) is most definitely *not* a science but rather an art.

9 In his seminal 1934 book *The logic of scientific discovery,* Karl Popper refers to his philosophy of science as "critical rationalism." For a more recent examination of his work, see J. Horgan, The intellectual warrior, *Scientific American,* Nov. 1992, 38–9.

10 The Kobe earthquake of January 17, 1995, has been examined extensively in the scientific literature. The reader may be interested in the following: Japan's seismic tragedy at Kobe, *Nature* (1995), 269; Geller, R. J., The role of seismology (correspondence), *Nature* (1995), 554; C. King, N. Koizumi, and Y. Kitagawa. Hydrogeochemical anomalies and the 1995 Kobe earthquake, *Science* (1995), 38–9; B. Johnstone, Complacency blamed for Kobe toll, *New Scientist,* Jan. 28, 1995, 4–5; P. Hadfield, Disaster quake wins grim place in record books, *New Scientist,* Feb. 18, 1995, 5.

3

Hazards of Shelter

San Francisco, 1906, and Messina, 1908

Both of these cities were struck by devastating earthquakes early in the morning, when most people were still in bed: San Francisco at 5:13 AM on April 18, 1906, and Messina at 5:23 AM on December 28, 1908. In more fundamental ways, these were similar geophysical events in two cities with similar geographical features, and the loss of personal property was comparable. The loss of human life, however, was drastically different.

A thriving seaport on the northeastern shore of Sicily, Messina had a population of some 150,000 when the earthquake struck. The death estimate published in April of 1909 was an unimaginable 100,000 in Messina alone and 50,000 in neighboring towns.[1] More recent sources do not modify these figures by much; the U.S. National Oceanic and Atmospheric Administration cites a total death toll of 120,000 in the wider event, of which perhaps 83,000 died in Messina proper.[2] By anyone's standards, this was a terrible disaster, with a survival rate of only 33% to 45% – in fact, this is probably one of the lowest human survival rates for any earthquake in history.

San Francisco's population in 1906 was 355,000, more than double that of 1908 Messina. Only one-fourth of San Francisco's structures escaped destruction by the earthquake and ensuing fire, which burned out of control for over three days.[3] Yet fewer than 700 people perished in the San Francisco disaster, yielding a survival rate of at least 99.8%.

At Messina, for every 1,000 residents, between 553 and 667 perished. At San Francisco, for every 1,000 residents, around 2 perished. In other words, an individual's chance of surviving the U.S. earthquake was some 280 to 480 times better than one's chance of surviving the Sicilian event of two years later.

Such a dramatic disparity in survival rates begs for an explanation. Was the Messina earthquake more intense? No, it turns out that the Messina event had a magnitude of 7.5 on the Richter Scale, whereas San Francisco experienced an 8.25. Because the Richter Scale is logarithmic rather than linear, it turns out that more than *5 times* as much seismic energy was released in the San Francisco event!

Tsunamis? Hardly a factor. The waves at Messina and surrounding towns were not particularly dramatic, they began with a trough rather than a more dangerous crest, and they did not pour into any major residential areas. At San Francisco, there apparently was no tsunami at all. Fire? San Francisco's was by far the worst and was brought under control only after the decision was made to dynamite several whole city blocks to provide firebreaks. At Messina, the few small fires were quickly extinguished by rain and cannot possibly account for any significant portion of the mind-boggling death toll.

What about the history of earthquakes in the two areas? Was one population more complacent than the other? This is a tougher call. At the time of the 1906 disaster, San Francisco could claim barely fifty years' worth of history as a major settlement, and only a small fraction of its population had been born in California. In 1856, when the city was only a mining town of small frame buildings, it experienced a violent shock that claimed a few victims, mainly from falling chimneys. An 1872 earthquake cracked walls in public buildings and caused a minor panic, but most of the few dozen fatalities resulted from a landslide outside of town. In 1898, several homes were destroyed and the navy shipyard suffered severe damage, but there was no loss of life. Yet throughout this period, quakes that could be felt averaged three or four a year. We can therefore assume that San Francisco's population of 1906 was generally familiar with the phenomenon of earthquakes but that very few had personally experienced a serious one.

Messina, by contrast, was hardly a boom town of recent immigrants. By 1908, this city could look back on more than two thousand years of recorded history, and the families of most of its residents had lived there continuously for many generations. Most knew that an official 29,515 had died in the catastrophic earthquake of February 5, 1783 (which destroyed most of the city), and virtually all had experienced the more recent destructive quakes of 1894, 1896, and September 8, 1905, the latter causing 529 deaths. The possibility of mass death and destruction from sudden earth movements was hardly an unfamiliar notion to those who lived in Messina in 1908. One can hardly argue, then, that the newly arrived residents of San Francisco were better informed of earthquake hazards than were the longtime residents of Messina.

Was geological science more advanced in California than in Sicily? No, again. A turn of the century geological map of southern Italy and Sicily,[4] showing fault lines and types of substrata, is at least as detailed and accurate as any contemporary maps of the San Francisco area. Moreover, if the California scientists could legitimately claim any credit for the high survival rate in San Francisco, then why was their academic home, Stanford University, so badly damaged in the event?

None of the factors just mentioned successfully explains the dramatic disparity in survival rates. To answer the question of why some 120,000 died at Messina while only 700 died in a stronger earthquake in the larger city of San Francisco, we need to look at the types of structures that housed the two populations as they were sleeping on those two fateful mornings. In San Francisco, most of the structures were wood, a cheap and plentiful building material for that rapidly growing city. In Messina, houses were predominantly masonry, with massive stone floors and brick-tile roofs supported by timbers set into niches in the granite walls. When Messina's walls began to wiggle in the earthquake, timber joists all over town slipped from their wall niches and allowed the heavy masonry above to crash onto the occupants below. The unsecured walls then fell in on top of the rubble. There wasn't much of a fire, because there was so little to burn; nearly everything was stone. In San Francisco, most of the timber buildings flexed resiliently in the tremor and survived relatively intact, as long as their foundations didn't shift. Most occupants therefore had plenty of time to get out well before the inevitable fire.

The message is clear: The death toll from an earthquake has more to do with the type of building construction than with the intensity of the earthquake. Earthquakes themselves seldom kill people; for the most part, it is our buildings that kill people.

Natural and Composite Materials

For several billions of years, the processes of geological and biological evolution have driven Mother Nature's development lab for structural design. When early humans arrived on the scene, trees had long since learned to sway in the wind yet keep their roots anchored in the earth. Stone offers a different kind of permanence: its resistance to fire and decay. In outlasting the human life span by a significant margin, timber and stone have long been obvious choices for fabricating shelters and other man-made structures.

Jacob Bronowski argues persuasively that before early humans could build, they first had to learn to take things apart.[5] This statement goes deeper than the obvious need to chop down saplings to construct a hut or to break stone into manageable chunks, to pile them up into a wall. Rather, it was essential that the early builders, through trial and error, study the con-

ditions under which natural materials fail. To process nature's raw materials
into useful construction materials, we need to induce failures in a controlled
and predictable manner; we must learn, in other words, how Mother Nature
will allow her handiwork to be disassembled. Then, to reassemble these
processed materials into new configurations, we must take care to *avoid*
those same conditions that will lead to materials failures. A shelter serves us
poorly if it is prone to unanticipated collapse.

The expectation that Mother Nature can be understood by taking her
handiwork apart and studying the properties of the pieces is generally
referred to as *scientific reductionism*. This intellectual process takes place at
the level of the base of the pyramid in Figure 2.2 of the last chapter, and its
objective is to discover those underlying principles that govern the behaviors
of the broadest possible variety of natural artifacts. The basic tenet of reductionism is this: If we know how materials behave at their most basic levels,
we should be able to predict how they will behave when we combine them in
new configurations that have never been observed in nature (brick houses,
for instance).

Clearly, there is a shortcoming in this line of thinking, for it considers the
whole of a structure to be no more than the sum of its parts. In practice,
there are always surprises – unanticipated consequences – any time natural
materials are reassembled in unnatural configurations. In fact, modern
materials scientists and engineers are well aware of this shortcoming, and
today very few are pure reductionists in their thinking. Still, when we seek
guidelines to the world of structural possibilities and caveats regarding
structural impossibilities, the knowledge that has been acquired through
reductionistic thinking continues to be quite useful.

Engineering is a tougher business than science. When a structure fails,
the blame falls upon the engineers and contractors rather than on the scientists who provided the theoretical underpinnings of their tangible work.
Often, this culpability is justified; we expect our structures to be designed
in accordance with scientific principles (as we currently know them) and the
building codes that embody these principles.[6] When an established scientific
paradigm has been ignored or applied incorrectly, it is the fault of the engineer or contractor. If the pilings of a beachfront building on Cape Hatteras
are undermined in a storm swell, for instance, we naturally question the
engineering. On the other hand, if a building in Boston should collapse in a
freak earthquake, we don't blame the scientists for having failed to anticipate
the earthquake hazard. Instead, we acknowledge that more scientific study is
needed regarding East Coast earthquakes.

Although such a limited social expectation may seem to be a blueprint for scientific irresponsibility, in fact this has not happened. Through the centuries, large numbers of physical scientists have always been diligent about struggling to uncover those principles which have direct social applications. In the next few sections, we'll explore some of the scientific principles that have proven useful in predicting how materials behave in structures.

Forces and Physical Events

The notion of a "force" takes root in our early childhood experiences: We push things, we pull things, and we observe that the things we push or pull sometimes move or break. Even the most mechanically inept of us learn at some level to be an engineer, seeking to reorder our physical surroundings to reflect our needs. But to know only how things "sometimes" respond to our efforts is a program for disappointment and frustration. When we entrust our lives and limbs to the unknown engineers who designed our buildings, dams, bridges, and elevators, we expect that their design calculations were based on something more substantial than "sometimes" connections with future behaviors of the materials.

By definition, a *force* is a push or a pull in a specific direction. To measure a force, one must observe its effects: A simple bathroom scale, for instance, measures the force pressing on the platform by linking the compression of an internal spring with the movement of a pointer on a dial. As I pointed out in Chapter 2, every measurement is the comparison of a physical quantity with some standard unit. Although the official international unit for measuring forces is the *newton* (N), in practice this official unit sees little use outside the scientific community.[7] Worldwide, the most common unit for expressing forces is the *kilogram-force* (kgf), a unit many scientists prefer not to acknowledge (although engineers are quite happy with it).[8] In the United States, the customary force unit is the *pound* (lb). By international agreement, these three force units are related as follows:

$$1 \text{ lb} = 4.448\ 222 \text{ N} = 0.453\ 592\ 37 \text{ kgf}$$
$$1 \text{ kgf} = 2.204\ 623 \text{ lb} = 9.806\ 65 \text{ N}$$
$$1 \text{ N} = 0.224\ 808\ 9 \text{ lb} = 0.101\ 971\ 6 \text{ kgf}$$

Of course, this degree of precision is not often used in engineering analysis, for it makes no sense to compute and document those digits that cannot in practice be measured. For readers in the United States, it's useful to remember that a kilogram-force is about 2.2 pounds, and that a newton is about a quarter pound (roughly the weight of a small apple).

It turns out that forces may or may not produce motion, and that motion may or may not require forces to sustain it. The relationships between forces and the behaviors of physical objects are considerably more subtle than that, as was first pointed out by Isaac Newton in the 1660s. Built into Newton's force laws are several underlying concepts.

First is the premise that forces are capable of canceling each other out if they act in opposing directions; thus, only the *unbalanced* portion of a collection of forces remains available to affect an object's motion. A disaster may subject a building to some very large forces, but if these are opposed by equally large forces from within the structure, the net unbalanced force will be zero, and the building will be happy to remain at rest. The challenge to structural engineers is to endow a structure with the ability to generate the internal forces that will counteract any external forces (snow loads, wind, etc.) that may arise during the building's lifetime. As we shall see shortly, this can be done by carefully choosing structural materials according to their elastic properties.

Second is the idea that motion is never absolute. You may think you're at rest as you read this book, but an equally valid point of view is that both you and the book are racing around Earth's axis at identical speeds (perhaps 1,000 kilometers per hour depending on your latitude). From yet another perspective, this motion of Earth on its axis is superimposed on the motion of Earth orbiting the Sun at some 41,000 kilometers per hour, so both you and the book are moving at approximately *that* speed. And so on, to the motion of the Sun and the motion of the entire Galaxy. You can't say how fast you're going in any absolute sense, because you can't point to something that is absolutely at rest to use as a reference. When you feel a breeze, it is not necessarily the air that is moving and you that are at rest; rather, the most you can say is that you and the air are moving at *different* speeds.

No force is needed to keep you (or the atmosphere) moving at a steady speed, regardless of how great that speed; it turns out that a force is needed only to *change* an object's speed, or to change the direction of its motion. A barge that is carried downstream on a river experiences no unbalanced force, as long as it rides with the current. But if a bridge pier lies in the barge's path, the impact will generate a significant and potentially damaging force on both objects. This collision force is the same regardless of which object (barge or pier) is "actually" moving prior to the impact.

A third and much more subtle idea, which I have already touched upon, is this: The distinction between cause and effect has no basis in nature. Forces always occur in equal and opposite pairs, and there are no objective criteria one can use to decide which is the cause and which is the effect. If you accidentally hit your finger with a hammer, the natural intellectual response (implicit in your cursing) is to view the hammer as the cause of a physical event whose effect is a bruised finger. Equally valid, however, is the observation that the *finger* caused an abrupt stop in the hammer's motion. One can view the hammer as the cause of the impact, or one can view the finger as the cause of the impact. Which is it, really? Clearly, there is no purely objective way to decide. All we can say objectively is that the physical event was characterized by a mutual interaction between the finger and the hammer.

Still, the language of cause and effect survives, because it is a simple way to reflect our human value judgments. If high winds buffet a house, we as humans are certainly more concerned about the effect of the wind on the house than the effect of the house on the wind. Newton tells us, however, that the physical analysis can proceed either way. This insight can be quite useful, for if we can analyze how a structure affects the wind (perhaps by studying a model in a wind tunnel), we also learn about the forces the wind reciprocally exerts on the structure.

In any literature, including scientific literature, the words "cause" and "effect" always imply that the writer is describing an event in terms of his or her own value judgments or priorities. I personally use the word "because" as much as anybody, and I make no apologies for doing so. Yet at their fundamental levels, the events of the Universe are driven only by mutual interactions, and there is no way unambiguously to distinguish "causes" from "effects." It is we, as humans, that attach more significance to the part of the mutual interaction that most concerns us.

Types of Forces

Newton's laws of motion describe the interactions between forces and material objects. These laws do not, however, tell us anything about how the forces themselves originate. To examine the behavior of structures, we first need to identify some specific types of forces.

Gravity: Newton recognized that every object in the universe attracts every other object in the universe through a gravitational force. In most cases, this is an extremely weak force; it takes a very massive object (the entire Planet Earth, for instance) to give rise to a significant gravitational force on human-sized objects. An object's *weight* is the gravitational force that attracts it to Earth. Obviously, all buildings need to be designed to support the weight of their structural members themselves, plus any weight added by human occupants, snow on the roof, and so on.

Gravity also accounts for the tides. The Moon's gravitational attraction distorts the world's oceans into a pair of tidal bulges that sweep around the globe as Earth rotates. Approximately twice each month, when Earth, Moon, and Sun are aligned, the tidal fluctuation (high tide to low tide) is particularly great. A typhoon striking a shoreline during such an alignment can be particularly devastating.

Friction: This is a force that acts to retard (and sometimes prevent) relative motion between two surfaces in contact. Friction is highly sensitive to the microscopic properties of the surface interfaces, to the presence of lubricants or contaminants, and to the force pressing the surfaces together. The surface contact area and the temperature are minor influences. Every structure depends on friction to at least some extent. Many older buildings simply rest on their foundations and depend on gravity and friction to keep them in place. Nails, staples, and nut-and-bolt assemblies also hold because of friction. As of this writing, our mathematical predictions of frictional forces are approximate at best; as a result, any structural connector that depends on friction should always be overdesigned by a considerable margin.

Fluid Static Forces: A fluid is any substance capable of flowing; for our purposes the important examples are the atmosphere, water, molten lava, and sometimes mud. Fluids, of course, transmit their weight to anything beneath them. They also, however, are capable of transmitting part of their weight in an upward direction, an effect known as *buoyancy.*

The Greek scientist Archimedes viewed it this way: When an object is immersed in a fluid, the object displaces a portion of the fluid and effectively creates a hole within the fluid. As the fluid attempts to flow into this hole, it pushes the immersed object upward, generating a buoyant force. Of course, the object may not actually *move* upward if there are other forces present to counteract the buoyancy, and it may even sink downward if the buoyancy is insufficient to support its weight. Nevertheless, objects in fluids do always experience some degree of buoyancy. A house in even a slow-moving floodwater may float away because of this effect.

Fluid Dynamic Forces: Fluids normally flow from regions of high pressure toward regions of lower pressure. Moving fluids obey Newton's law of inertia: once moving, they tend to continue moving in a straight line. If a moving fluid is deflected or stopped by an obstacle, it can transmit a very large inertial force to that obstacle. Rapidly moving water, for example, has no trouble uprooting trees and sweeping away buildings (Fig. 3.1).

In addition to inertial forces, moving fluids can also generate dynamic lift forces. The phenomenon of lift is routinely exploited in the design of both airplane wings and the hulls of hydroplaning boats. The effect is easily demonstrated by holding a sheet of paper horizontally in front of your mouth and blowing over the upper surface. The paper will rise into the fast-moving air. In the same way, the roof of a house will experience a lifting force in a high wind.

Fluid Friction: A moving fluid transmits a frictional force to the sides of any channel it flows through. The faster the movement, the greater

Figure 3.1. Testimony to the power of moving water. The wreckage of the John Schultz house on Union Street in Johnstown, after the flood of 1889. (Journalist, Pittsburgh, 1889.)

this force. For example, a small hole in a dam tends to quickly erode into a large hole and can lead to the failure of the entire structure. Because of fluid friction, an object carried away by a moving fluid will quickly attain the same speed as the fluid. For this same reason, something as seemingly innocuous as an unsecured porch chair or even a pine needle can become a hazardous projectile in a hurricane.

Elastic Forces: Every solid exhibits the ability to "remember" its original shape when it is deformed slightly by external forces. When the external force is removed, elastic forces within the solid return it to its original shape. A tree will bend in a breeze, for instance, then will spring back when the wind calms down. Of course, there is a limit to this behavior, and exceeding the limit will cause a failure. Structural materials are chosen for their ability to generate such internal elastic forces throughout the range of expected deformations.

Strengths of Materials

Around 1670, the English natural philosopher Robert Hooke did a series of studies on the properties of springs. He found that if a particular spring required, say, a 2-kgf force to stretch it by 1 cm, then a force of 4 kgf would be needed to stretch the same spring by 2 cm and 6 kgf for 3 cm. The amount of deformation, in other words, was directly proportional to the amount of force. Stiff springs would carry larger forces than limp springs, but in every case doubling the force would double the deformation, and tripling the force would triple the deformation. Moreover, springs repeatedly return to their original shape when the deforming force is removed (which, of course, is what allows them to be springs).

Hooke also noted that this behavior holds true only up to a limit. If you apply a large enough force to a spring, it will stretch considerably farther than is predicted by a simple proportional relationship. And then, when you remove this large force, the spring will not return completely to its original shape. This observation is summarized by saying that every spring has an *elastic limit.* Unfortunately, to measure the elastic limit you need to effectively destroy the spring.

Why am I talking about springs? Because it turns out that all structural materials – whether stone, concrete, timber, or steel beams – behave like very stiff springs. Structural materials need to be elastic; otherwise they can't carry a load. It's quite impossible to build even a small structure from putty or soft modeling clay – materials that lack elasticity.

Clearly, a steel beam does not deform a great deal under a load. It does, however, deform a little, and this deformation can be measured with proper testing equipment. The elastic limit of a steel beam is also measurable, although only through heroic efforts that necessarily destroy the beam. Of course, if we want to know how much weight a beam will support, it does us little good to destroy the beam to get the information. What we need to do is *predict* the load the beam can carry.

We begin by observing that there are several ways to deform a solid material (Fig. 3.2). The most common type of deformation is a *compression*, as when a load acts to squash a supporting column. If a structural member is being stretched (a steel cable, for instance), we say it is in *tension*. *Bending*, as we will see shortly, can be considered as a combination of tension and compression in different parts of the member. The design of fasteners must also consider *torsion*, which is a twisting deformation, as well as *shear*, which is the relative displacement along a pair of parallel planes within the material.

When Mother Nature pummels a building, the deformations most critical to the building's survival are tension, compression, and bending. Although a complete mathematical analysis of these deformations can become quite complex, the fundamental principles can be described with some relatively simple arithmetic, as follows.

Clearly, a thick structural member will be stronger than a thin one of the same material. In tension and compression, the length of the member doesn't affect its strength: if a 10-meter length of rope will lift 200 kgf, a 5-

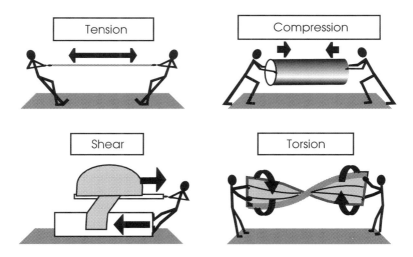

Figure 3.2. The modes of deformation of solid materials.

meter length of the same rope will lift no more and no less. In determining
the strength of a structural member, the important geometrical factor is the
cross-sectional area that supports the load. The ratio of the load force to this
supporting area is called the *stress:*

$$\text{Stress} = \frac{\text{Load force}}{\text{Supporting area}}.$$

If the load force is expressed in units of kgf and the supporting area is in
square centimeters, the stress has units of *kilogram-force per square centimeter* (kgf/cm²). In the United States, a common alternative unit is the *pound
per square inch* (lb/in², or psi). For instance, in Figure 3.3 we see a column
supporting a load force of 12,000 kgf. The supporting area is 9 cm by 9 cm,
or 81 cm². The stress on this column is the applied weight divided by the
area, or 148 kgf/cm². The column's height has nothing to do with this result.

   Standing alone, this computation is not particularly interesting or useful.
However, if we look at Table 3.1, we see that some materials can safely sup-

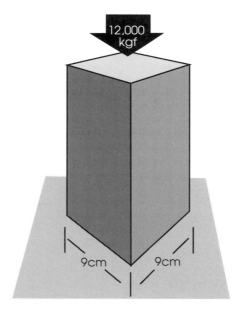

Figure 3.3. A vertical column supporting a compressive load of 12,000 kgf.
Given the cross-sectional area of 81 cm², the compressive stress is 148 kgf/cm².
If this column is made of concrete or brick, it will fail. If it is made of steel, it
will support this load.

port our calculated 148 kgf per square centimeter, and other materials cannot. If the column in Figure 3.3 were built of common brick or concrete, it would immediately fail when our load was applied. On the other hand, a solid steel column would be safe by a considerable margin, since structural steel can withstand up to 2,500 kgf per square centimeter without exceeding its elastic limit.

Does this mean that it's impossible to support a 12,000-kgf weight with a concrete column? Not at all. It simply means that a concrete column would need to be beefed up beyond the 9 cm by 9 cm size we used in this example.

Table 3.1. *The elastic limits and ultimate strengths of some representative materials* (kgf/cm^2)

Solid	Elastic limit		Ultimate strength	
	Tension	Compression	Tension	Compression
Aluminum	840	840	1,800	Indefinite
Brick, best hard	30	840	30	840
Brick, common	4	70	4	70
Concrete, Portland, 1 mo old	14	70	14	70
Concrete, Portland, 1 yr old	28	140	28	140
Douglas fir	330	330	500	430
Granite	49	1,300	49	1,300
Iron, cast	420	1,800	1,400	5,600
Limestone and sandstone	21	630	21	630
Oak, white	310	310	600	520
Pine, white	270	270	400	345
Slate	35	980	35	980
Steel bridge cable	6,700	6,700	15,000	15,000
Steel, 1% carbon, tempered	5,000	5,000	8,400	8,400
Steel, chrome, tempered	9,100	9,100	11,000	11,000
Steel, stainless	2,100	2,100	5,300	5,300
Steel, structural	2,500	2,500	4,600	4,600

Note: 1 kgf/cm^2 = 14.22 lb/in^2. Adapted from E. Zebrowski, Jr., *Practical physics* (McGraw-Hill, 1980.)

Suppose, for instance, that we proposed to support the 12,000 kgf on a con-crete column having cross-sectional dimensions of 23 cm by 23 cm. Now the supporting area is 529 cm^2 and the stress is 12,000 kgf/529 cm^2, or 22.7 kgf/cm^2. Referring to Table 3.1, we see that the elastic limit of even rela-tively young concrete is 70 kgf/cm^2. In this case, then, we would be stress-ing the material to a bit less than one-third of its compressional elastic limit, and a concrete column could be expected to perform quite adequately.

Notice, from Table 3.1, that some materials have different elastic limit in tension and in compression; engineers must therefore anticipate whether their chosen materials will be subject to tensile or compressional stresses after they have been integrated into a complex structure. This assessment may sometimes be difficult to do, particularly for structures that could pos-sibly experience a variety of transient forces associated with high winds, earthquakes, and the like.

Notice also that each material is characterized by an "ultimate strength," which may or may not be the same as its elastic limit. The distinction can be seen in the graph in Figure 3.4, which represents the general relationship between the stress applied to a material sample and the sample's deforma-

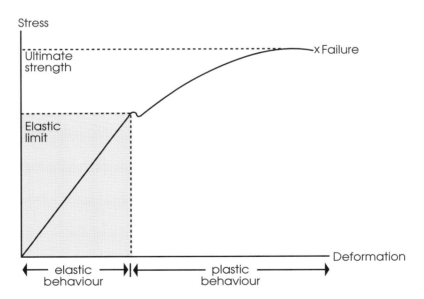

Figure 3.4. The relationship between stress and deformation for a typical solid material. At low stresses, most solids behave elastically. When the stress exceeds the elastic limit, many materials exhibit plastic behaviour before they fail.

tion (i.e., the amount by which it stretches or compresses). At low stresses, regardless of the material, the graph begins as a straight line; this is the region of *elastic behavior*, where the solid deforms under stress but returns to its original shape when the stress is removed. The elastic limit is the largest stress one can apply before inducing a permanent deformation in the material. Some materials (e.g., brick) break apart immediately when the elastic limit is exceeded, but other materials (e.g., wood or steel) enter a region of *plastic behavior* where they undergo a permanent deformation without actually breaking. One can easily demonstrate this with a coat hanger or paper clip; bend it a little and it pops back, but bend it too far and it retains the new shape you've imposed. Eventually, however, given enough stress, every material will break. For practical purposes, the *ultimate strength* is the stress that will induce a total failure.

In structures, it is essential that none of the materials experience stresses beyond their elastic limits. In fact, to allow for the contingencies of unanticipated loads, deterioration over time, and so on, engineering standards usually require that the design stresses be kept to less than one-third of the elastic limits of the materials.

What happens if loads arise on a structure that push some of the materials beyond their elastic limits? If the key materials have no plastic range (stone, for instance), the structure will collapse immediately into a pile of rubble. On the other hand, if the stresses stay within the plastic ranges of the key structural members, the building will be left leaning, sagging, or twisted out of kilter, but still with its parts all joined together. Such buildings will usually need to be demolished later, but during the life-threatening period of a disaster they will have provided their occupants an increased chance of survival. It is this plastic behavior of wood (versus stone) that explains the dramatic difference in survival rates in the San Francisco and Messina earthquakes of 1906 and 1908.

Bending

The majority of structural failures result from the excessive bending of a critical structural member. Figure 3.5 shows a horizontal beam supported at its ends and carrying a vertical load near its center. As this beam deflects, the material near the concave surface (the inside of the bend) goes into compression while the material near the convex surface goes into tension. You can see, and indeed feel, this effect if you bend a stick across your knee.

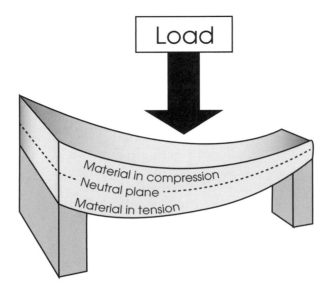

Figure 3.5. The bending of a beam. The material on the concave side of the bend goes into compression, while the material on the convex side goes into tension.

Clearly, if one layer of the material is in compression and a parallel layer is in tension, there must be an intermediate layer where the material experiences no force. This interface between the tension side and the compression side is called the *neutral plane*. The material near the neutral plane plays virtually no role in contributing to the strength of the beam; in fact, if you want to drill a hole in the beam to run a pipe or electrical wire, this is the place to do it. On the other hand, the points farthest from the neutral plane must carry the largest tensile and compressive forces. For this reason, beams are often designed as in Figure 3.6, with the largest cross-sections of material located as far as practical from the neutral plane. Meanwhile, openings may purposely be introduced near the neutral plane, because this reduces the beam's weight without reducing its strength (the removed material would have been loafing anyway). We also notice in this diagram that a beam can be constructed from two parallel slabs of material connected by a series of light triangles. Technically, this arrangement is called a *truss*, but engineers often view it as a beam with the unimportant materials removed.

Beams are an essential part of most structures, from houses to bridges to skyscrapers (which can be thought of as great vertical beams with one end

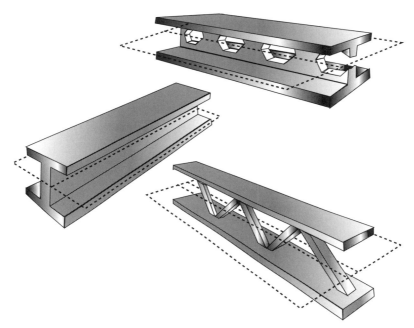

Figure 3.6. The design of beams. A beam's strength-to-weight ratio is improved by locating most of the material as far as possible from the neutral plane.

stuck in the ground). Clearly, the load-carrying ability of a beam depends on its geometry and on how the load is applied. More importantly, though, the strength of a beam will depend on the material it is made from. Even the best-designed stone beam will function poorly compared to an inefficiently designed steel beam.

Referring back to Table 3.1, we see that most woods and metals have roughly equal elastic limits whether in tension or compression. Such materials are appropriate for making beams, because, when subjected to bending, such a beam's tension and compression sides are equally strong. On the other hand, materials like cast iron, stone, brick, and concrete are much stronger in compression than in tension (by factors ranging from 4 to 30). Beams made of these latter materials are limited in strength, due to the weakness of the tension side when they bend. Suppose, for instance, that the horizontal beam of Figure 3.5 is made of limestone. This beam will fail when the material near the bottom surface experiences a tension of 21 kgf per square centimeter of cross-section. Meanwhile, the material near the top is loafing. (Although the actual stress here is about the same as that at the bottom, the top material is capable of carrying some 630 kgf of compression per square cm, or some 30 times as much.) The considerable compressive strength of the limestone has been wasted, insofar as it contributes nothing to the strength of the beam as a whole.

One might argue that this isn't a serious drawback, because one can simply make the beam thicker and taller – that is, increase the cross-sectional area and thereby the strength. Such logic, however, fails to account for another variable: the weight of the structural material itself. The material that's loafing on the compression side isn't inconsequential: It is contributing a considerable weight to the beam. This weight adds to the stresses that must be carried by the beam's weaker tension side. Beyond a certain point, making a masonry beam thicker makes it no stronger at all.

In spite of this drawback, horizontal stone beams were used widely in antiquity, most notably by the Greeks. If you visit the ruins of such ancient buildings today, you commonly find the vertical columns still standing but the original horizontal lintels and roofs reduced to pieces of rubble strewn about the ground (Fig. 3.7). Were the ancient builders unaware of the structural shortcomings of stone when used for horizontal beams? That is unlikely. After all, shaping stone for building purposes depends on an ability to induce predictable failures. Figure 3.8 shows how a mason cleaves a small stone or brick. The method is based on the idea, in modern language, of inducing a tension failure. To shape stone through compressional failures is some 30 to 40 times as difficult, and the ancient Greek builders surely were aware of this.

To compensate for the low tensile strength of stone, the Greek builders limited their use of horizontal stone beams to relatively short spans. As a result, Greek public architecture came to be characterized by large numbers of closely spaced supporting columns, both on the perimeter and in the interior of each building. (Vast open interior spaces were probably well beyond their imagination at that time.) The compressional load-bearing capacity of a vertical supporting column is not a limiting factor, for in nature we see beds of stone supporting whole mountains. Horizontal stone beams, however, are *not* found in nature, so here one needs to be careful. And the Greeks indeed were careful. Most of their buildings collapsed only because of events that lay well beyond their ability to plan for. In most cases, the failures were due to earthquakes that overstressed the tension side of the stone lintels supporting the roofs. Today, one can still wander through these ruins and marvel at the sight of what is left. Most of the structural elements that were originally designed to carry only compressive stresses continue to stand, more than twenty-four hundred years after their erection.

Figure 3.7. Ancient Greek ruins at Corinth. Stone is vulnerable to failure if permitted to go into tension. This structure was destroyed by an earthquake in A.D. 856 that claimed approximately 45,000 lives. (Photo by author.)

Modern poured concrete has properties very similar to those of the stone used by the ancients; it is strong in compression but considerably weaker in tension. How, then, do engineers get away with using concrete in the horizontal roadways on bridges, or the floors of high-rise apartment buildings, or in similar applications where the material will be subject to bending stresses? The answer is that the concrete in such applications does not stand alone. A reliable concrete beam demands that the tension side be reinforced with imbedded steel rods or wire mesh, which provide the necessary tensile strength. If the beam in Figure 3.5 were made of concrete, we would clearly want to imbed steel wire or rods near the bottom as we poured the concrete. (To put the steel at or above the neutral plane would be totally ineffective.) Concrete which has been properly reinforced with steel rods or mesh on its tension side is referred to as *reinforced concrete*. The strength of this material is improved even more by stretching the steel reinforcing rods and hold-

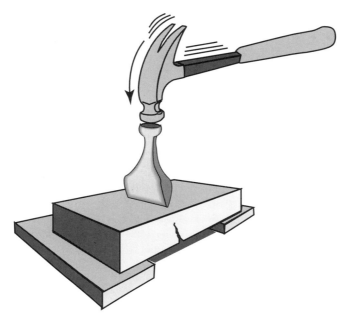

Figure 3.8. Cleaving a brick. The failure begins at the tension face of the brick and travels upward to the relatively blunt chisel. This is considerably easier than attempting to induce a compression failure.

ing them in tension while pouring and curing the surrounding concrete. The resulting material is called *prestressed concrete.*

The practice of using unreinforced masonry for vertical supporting columns has had a long and generally successful history. Such columns do, however, fail when they are subjected to large horizontal bending forces: the lateral shear of earthquakes, the horizontal action of ocean waves, or the fast-moving currents of floods, for instance. In regions where such events are anticipated, vertical prestressed concrete columns can make a major contribution to the integrity of a structure. In some parts of the southern and western United States, utility poles are now being made of prestressed concrete; such poles are unlikely to be toppled by any imaginable natural disaster.

Triangles and Arches

In designing a structure, there is a critical interplay between the properties of the materials chosen and the geometrical configuration in which they will be expected to function. Two particularly important structural geometries are triangles and arches.

The triangle is the only inherently rigid geometrical shape. By this, we mean that a triangle cannot be distorted without breaking it apart. On the other hand, it is quite easy to distort a rectangle into a parallelogram, a hexagon into a flattened hexagon, and so on with other polygons, without breaking these shapes apart. If you want to stiffen up a rectangle, the way to do it is to introduce triangles. This principle is used even where a physical triangle may not be apparent at first glance; in nailing two boards together, for instance, it is much more effective to arrange the nails in triangular patterns than to run them all in a straight line.

In Figure 3.9, we use this principle to reduce the deflection of a long horizontal beam. Notice that the sides of the triangles must be pinned together where they meet, but that it is not necessary to secure these pins against rotation. There is simply no way any part of these triangles *can* rotate, as long as the structural members themselves remain intact. For this reason, construction codes in earthquake and hurricane-prone areas specify that

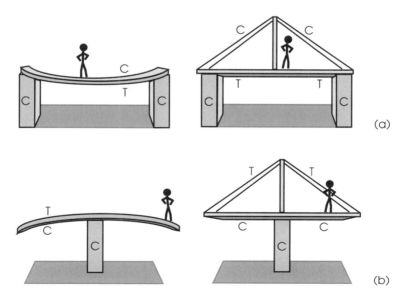

Figure 3.9. Using triangles to reduce the deflection of a horizontal beam: (*a*) a simple beam supported at its ends and the same beam in a truss; (*b*) a simple beam supported at its center and the same beam in a balanced cantilever. Tension members are label *T* and compression members *C*

houses include triangular bracing in their frames to stiffen the structure against twisting motions. This is relatively simple thing to do.

However, one other issue needs to be examined: Which sides of the triangles are in tension, and which are in compression? If a member is expected to carry a tension, we can fabricate it from a steel rod or even a flexible cable. But if a member is to carry a compression, a cable won't do at all (it would just go slack, and the structure would quickly collapse). The choice of material for each piece, as well as the choice of fasteners, must reflect whether the piece is to carry a tensile or compressional stress.

Fortunately, predicting which pieces are in tension and which are in compression really isn't too difficult. In Figure 3.9a, we see that the vertical post in the center must be a tension member, since it pulls up on the middle of the beam. At the top, then, this same member must be pulling down (equal and opposite action and reaction), and it therefore compresses the two diagonal pieces. These diagonals, in turn, push down and outward against the piers. The original beam then becomes a tension member, since it keeps the lower corners of the triangles from moving apart. The piers, of course, are in compression. The configuration in Figure 3.9b may superficially appear to be geometrically similar, but structurally it is quite different. Here the diagonals of the triangles are tension members, the central post is a compression member, and the original beam becomes a compression member once it's been pulled horizontal.

Notice that the introduction of triangles to stiffen a structure always introduces new tensile and compressional forces in the structure. These forces of tension and compression must generally oppose each other. Moreover, the materials used must have enough elastic stiffness so that when Mother Nature introduces a new external force (say a high wind or an earthquake tremor) the original elastic forces can readjust themselves without a large amount of structural movement. It is not enough that our buildings stand up; we also want them to stand without moving and shaking a great deal.

Although the triangle is the only inherently rigid geometrical shape, one other shape can be *made* rigid if it is loaded properly: the arch. Scholars still argue whether the invention of the arch should be credited to the ancient Romans or to earlier civilizations. This is a moot point, for it was the Romans who, in first using the arch in large-scale structures, demonstrated their confidence in predicting the long-term behavior of this structural element. Many Roman arched aqueducts, bridges, and sewers are still functional after twenty centuries of exposure to the forces of nature; one of many grand examples is the Pont-du-Gard at Nimes (in southern France), which spans a 250-meter (800-ft) valley with three tiers of arches whose height totals 47 meters (155 ft). The Pont-du-Gard was built in A.D. 14 using cut stone and no mortar. Although it no longer carries water (its original function), it is still used today as a footbridge.

As shown in Figure 3.10, Roman arches were semicircular and built of unreinforced masonry. Although mortar was sometimes used between the stones, this was no more than a grout of lime and volcanic ash, which had virtually no tensile strength and therefore had little effect on the overall integrity of the structure. Rather, the arch derives its strength from a wonderful combination of three physical factors: (1) stone is very heavy, (2) the weight of a properly shaped piece of stone transfers a compressive stress to the stones beneath it, and (3) stone is very strong under compressive stresses. If an arch is designed so that no part of it ever goes into tension, the structure will stand for thousands of years, until the masonry material itself begins eroding into dust.

The design of an arch is quite simple; just cut each of the arch stones so it is a little bit wedge shaped, with the widest part on top. Then no stone in a continuous semicircle can slip through, and each keeps its neighbors in place. Actual construction, however, is somewhat more challenging, because the arch does not become self-supporting until the last stone is in place. This last stone, often referred to as the keystone, is usually the center stone. Until the keystone is set, the arch must be supported on temporary timber falsework. The practical requirement of assembling this falsework limited the distance the Romans would attempt to span with a single arch: It was much

Figure 3.10. A series of Roman arches. To prevent collapse, every part of each arch must remain in compression.

easier (and certainly safer during construction) to bridge a valley with a series of smaller arches than with a single sweeping arch. For this reason, we see Roman construction characterized by amazing numbers of sequential arches, often supporting more arches above them, and sometimes topped by yet additional tiers of arches.

The load carried above the center of an arch transmits an outward thrust to its supporting columns; if such a force causes a column to bend even a little, the arch will quickly collapse. In a series of identical arches, each braced against the next, this creates a problem only with the last arch at each end. Of course, if the end arch does collapse, the neighboring arch becomes the new end arch, and a failure will proceed domino fashion through the structure. But here again the Romans had creative solutions. In their aqueducts and bridges, the last arch on each end of the structure was always built to transmit its outward thrust against bedrock (Fig. 3.10). In the cities, arched structures were either built with extremely massive vertical columns (as in Rome's Arch of Titus, built in A.D. 81 and still standing), or else by using a circular or oval floor plan (as in the Coliseum in Rome).[10] Circles and ovals have no stray ends.

A thousand years after the Romans, the builders of the Gothic cathedrals of western Europe made a vast improvement on the arch. The creative challenge they faced was that of increasing the height and openness of the interior space while simultaneously maximizing the size of the windows. Because window glass cannot support a vertical load, walls with big windows do not contribute much to the support of a structure. These builders' innovation was the Gothic arch, high and steep and supported on narrow vertical columns, with "flying buttresses" to support the outward thrust of the structure's walls and terminal arches. Even modern-day tourists who step into a medieval Gothic cathedral are usually struck with awe at how *little* stone was used in these magnificent buildings, relative to their vast open interior space. Clearly, the medieval architects were very good scientists, for they understood and successfully predicted that in every wall and arch, and in every masonry joint in this immense cathedral, every single piece of stone would continuously remain in a state of compression indefinitely into the future. And so indeed have these many thousands of pieces of stone performed, never experiencing a tension force, for the past eight hundred years.

The Gothic arch is higher than it is wide, but it can be built as wide as you wish if you are willing to make it high enough. By comparison, the low Roman arch was extremely wasteful in its use of materials, relative to the space it enclosed and relative to the maximum compressive load it could support at its weakest point (its center). Gothic architecture was clearly superior in these respects, and it spread throughout medieval Europe, even though construction was extremely time-consuming (often taking up to eighty years to build a single large church). The original Gothic buildings that survive today still represent the limits of what is structurally possible when using only unreinforced masonry.

Unfortunately, one cannot depend on stone structures, regardless of their elegance, in the face of an earthquake. In fact, if unreinforced masonry arches are light (i.e., incorporating Gothic rather than Roman geometry), one can actually be *less* confident of their surviving a major geophysical event. Only one stone needs to slip from an arch to initiate a complete collapse. When there is less weight holding the arch stones in compression, a lateral jostle is more likely to initiate a tension failure in a joint and the loss of a stone. Many of the deaths in Lisbon's terrible 1755 earthquake (discussed in Chapter 1) resulted from such collapses of unreinforced masonry Gothic arches and vaults in the churches and cathedrals.

Although the arch principle has been used in many magnificent bridges (such as the New River Gorge Bridge in West Virginia, which is supported by a steel arch spanning some 518 meters [1,700 ft]), the very largest and strongest arches are those built to lie on their sides and hold back huge reservoirs of water. The widest arched dams exceed 1 kilometer in width, the highest rise as high as 300 meters, and one in Canada impounds 142 billion cubic meters of water (the equivalent of some 10% of the volume of water in Lake Erie). The structural demands on dams exceed those on any other man-made structures by several orders of magnitude, and the potential human consequences of the failure of a large dam can boggle the imagination. The fact that modern dams are so reliable is testimony that many of the lessons passed on by the Romans are still taken seriously by today's civil engineers.

Johnstown, Pennsylvania, 1889

In May of 1889, record rains drenched the partially deforested Conemaugh River valley upstream of the small industrial city of Johnstown, population 28,000. Had the Conemaugh been a free-running river, it would have crested high enough to flood a portion of Johnstown's business district, damage the lower stories of a few hundred houses, render several bridges impassable, and interrupt some public services and utilities. Few if any lives would have been lost, for floods in this region usually gave observant people ample time to scamper to higher ground. In fact, Johnstown residents had long learned to expect annoying floods almost every spring, for the Connemaugh and its tributaries had flooded at particularly high levels in 1808, 1847, 1875, 1880, 1885, 1887, and 1888. The flood of 1889 would have been but slightly worse than these earlier floods if the hand of Man hadn't transformed the event into a major disaster. The aggravating factor was a dam that did not do its job.

This was the most devastating dam failure in U.S. history, and it involved a structure that the ancient Romans would have been smart enough *not* to build.[11] The dam was not arched, it did not transmit its load to bedrock, and it was not even built of masonry. The South Fork Dam was originally built to supply water to a section of the Pennsylvania Canal at a location on the South Fork of the Conemaugh River 22 kilometers (14 mi) upstream from Johnstown. This dam was an earthen structure 260 meters (850 ft) in width and 24 meters (80 ft) in height, which impounded a lake some 5 kilometers in length and around 2 kilometers in width. The dam originally failed in 1847, while it was still under construction, and on that occasion the resulting torrent washed out an aqueduct at Johnstown. Completed in 1852, the dam's original function was rendered obsolete within a decade as railroads replaced the canals. By 1889, the structure's fundamental design shortcomings had been aggravated by neglectful maintenance during ten years of private ownership by the South Fork Hunting and Fishing Club, an exclusive group of Pittsburgh millionaires. The spillway had been obstructed by a meshwork designed to keep fish from escaping, pipes originally installed to permit independent regulation of the water level had been removed (a cheaper strategy than repairing and maintaining them), and the breast of the dam had been allowed to sink lower than its shoulders. The failure was predictable, not in terms of the exact date and time, but in view of the statistical certainty that sooner or later the region was bound to be drenched by heavy rains whose runoff would exceed the capacity of the dam's spillway.

When a downpour began on May 30, 1889, it had already rained eleven days that month. Because the saturated earth of the surrounding forests could hold no more water, this rain ran off into the creeks and rivers. Around noon on the thirtieth, many of the more observant inhabitants of Johnstown and upstream villages began to evacuate the low-lying areas; had they not, the death toll would have been much higher. Upstream, the inadequate and obstructed spillway of the South Fork Dam became clogged with flood debris, and the waters of the artificial lake rose until they topped the breast of the dam. A few minutes after 3 PM on May 31, the dam exploded and sent a thundering wave front 15 meters (50 ft) high down the already-swollen river. It took around 36 minutes for the lake to empty, and for those 36 minutes fluid energy poured into the valley below at a rate comparable to the energy flow over Niagara Falls.

Within its first moments, the giant wave ripped up thousands of trees and churned them into a thundering maelstrom of splintered timber. The leading edge of this surge chewed through the forests and villages farther downstream. On the steeper straightaways of the valley, the wave reached speeds of 100 kilometers per hour (60 mi/h), while on sharp curves and shallower drops it slowed to around 15 kilometers per hour (10 mi/h). (According to some observers, it may almost have stopped at times.) Because the bottom of the surge was retarded by friction while the top was not, this great flood wave progressed not as a wall of water but rather as a turbulent, cascading breaker. Victims were pounded downward rather than being swept forward and upward. This was not the kind of flood which would allow anyone to swim to safety.

A few miles downstream stood the Connemaugh Viaduct, a large and substantial stone arch bridge. Here the debris jammed up, temporarily creating a leaky dam 22 meters (71 ft) high, which was about the same height as the dam that had burst. Had the viaduct held at this point, the downstream towns might have experienced only a modest acceleration in the rising water. Alas, although the viaduct's engineers had planned for floods in their design, they had not anticipated this particular extreme set of conditions. The viaduct held for only a few minutes. As the structure toppled, the flood wave surged forward with renewed vigor and completely wiped away the village of Mineral Point, destroying 30 houses and a furniture factory. It then burst across an oxbow, resculpturing the river's channel, and tore into East Connemagh, where it carried away whole railroad trains and thirty steam locomotives. Next downstream was Woodvale, where 255 houses were swept completely from their foundations, along with a tannery and a streetcar shed housing eighty-nine horses. When the water hit the Gautier Wireworks, boilers and furnaces promptly exploded and contributed a billowing "death cloud" of soot and ash to the churning wave front, along with many miles of tangled barbed wire.

As it thundered into Johnstown, this tumbling mass of debris indiscriminately ground up and swept away most human constructions in its path. Bodies of humans and animals churned in the foam, and even railroad cars were tossed around like footballs. In just a few terrible minutes, thousands of houses and businesses were destroyed, and thousands lost their lives. Ultimately, the official statistics would list 2,209 known dead and 967 missing. If you walk through Johnstown's cemeteries today, you won't fail to notice a particular recurring date on the tombstones: "Died May 31, 1889." One section of Grandview Cemetery has a large granite monument dedicated to the unknown dead, behind which are 777 small white and anonymous marble headstones.

Only one man-made structure survived the direct onslaught of the wave, and paradoxically its survival actually contributed to the horror of the disaster. This structure, a railroad bridge at the lower end of town, had been built as a series of seven low semicircular stone arches that would have looked quite at home in ancient Rome. When the thundering surge threw its debris against this bridge, it once again created a dam, just as it had at the upstream viaduct. Here, however, the individual arches were much shorter in span and heavier; this was a very conservatively built structure (which some engineers would characterize as inefficient in its use of materials), and it held. Although the Cambria Iron Company and many houses downstream of the Stone Bridge were damaged by the portion of the surge that topped the bridge, relatively few of these structures were totally destroyed.

Unfortunately, although Stone Bridge saved some, it also aggravated the disaster for many others. The sudden blockage of a rapidly moving stream always has its repercussions; on a much smaller scale you may have noticed the "water hammer" effect when the solenoid valve on a washing machine suddenly closes. When the arches of the Stone Bridge were slammed shut by the debris riding on the wave front, the flood wave was reflected back upriver and into a nearby tributary, Stony Creek, where it swept away even more dwellings in the town of Kernville. In places, this reflected wave may have attained a height of as much as 30 meters (100 ft), roughly double that of the initial wave.

Yet there was more to come. Upstream of the Stone Bridge the river was blanketed with floating trees, the remnants of crushed houses, and many partially intact structures that had been carried off their foundations. As the floodwaters poured over and through the logjam at the bridge, the trapped flotsam piled higher and higher. Some victims apparently did succeed in climbing out of this heap of debris. But then this giant pile of splintered wood caught fire, possibly ignited by an overturned stove in an upper story of one of the many houses that had ridden the floodwaters into the bridge, and possibly fueled further by oil leaking from wrecked railroad tank cars.

In spite of the continuing rain, there was enough dry fuel in this massive pileup to sustain a monstrous fire, which continued unchecked for several days (Fig. 3.11). It has been speculated that many of the missing – those whose bodies were never recovered – were consumed in this grisly conflagration at the Stone Bridge. After the floodwaters subsided and the flames finally died out, cleanup workers had to resort to explosives to clear the tangled and charred wreckage at the bridge in order to restore the course of the stream.

Relief poured in from across the country, but Pittsburgh's "Captains of Industry" made relatively token contributions; Andrew Carnegie, a member of the club that owned the dam, contributed just ten thousand dollars on behalf of his steel company, while thirty of the sixty-one club members contributed nothing at all. Ultimately, the courts in Pittsburgh ruled that the disaster had been an act of God, and that the South Fork Hunting and Fishing Club and its members had no legal liability for the damage or loss of life. Particularly galling to the survivors, many of whom had lost everything they owned and loved, was the fact that the dam had served no functional purpose other than to provide recreation for an exclusive private membership. This angry sentiment was captured in a widely quoted contemporary poem by one Isaac Reed:

> Many thousand human lives –
> Butchered husbands, slaughtered wives,
> > Mangled daughters, bleeding sons,
> > Hosts of martyred little ones,
> (Worse than Herod's awful crime)
> Sent to heaven before their time;
> > Lovers burnt and sweethearts drowned,
> > Darlings lost but never found!
> All the horrors that hell could wish,
> Such was the price that was paid for – fish!

The dam was never rebuilt, and the club's property was sold off. Today a few of the original club members' houses still stand, privately owned, and one can dine at a restaurant in the former clubhouse, which is now owned by a local historical society. The National Park Service maintains a memorial park and a museum that includes the site of the failed dam, the shoulders of which still remain.

The Johnstown Flood offered a profound lesson: Knowledge by itself is no guarantee of sound engineering practice. In 1889, all knowledgeable par-

Figure 3.11. The disaster at Johnstown's Stone Bridge, as depicted in an 1889 lithograph.

ties were quite aware that the South Fork Dam was unsafe, and that even the ancient technology of the Romans would have created a safer structure (as was subsequently confirmed by the survival of Johnstown's Stone Bridge). But knowledge alone does not guarantee the power to act, or to force others to act. The South Fork Dam was no more than a weekend recreational asset to its owners, and, by being part of a collective, the club members were largely insulated from the concerns that had been voiced over the years by various downstream residents. Members of the club were not necessarily evil people (although some of the officers certainly had misplaced priorities); a better characterization is that most of the members were ignorant of the physical principles, ignorant of the prevailing engineering practices, and/or too distracted with other activities to contemplate the potential human consequences of a failure of their dam. They were mainly uninformed, and, through their insularity, uninformable.

Today, one can no longer build or own a private dam on a public waterway. In the United States, to dam even a small stream on private property requires environmental impact studies, hearings, licenses, engineering reviews, and subsequent inspections. This body of governmental regulation has evolved in response to the historical record of incidents where individuals and private groups had been presumed to be safeguarding the public interest but in fact failed to do so. Those who complain about today's "excessive" government regulation lose all credibility if they ignore the historical disasters that were aggravated by the lack of such regulation. The Johnstown Flood of 1889 stands as a prime example, not to be forgotten.

Joints, Fasteners, and Foundations

The strength of the materials alone does not guarantee the integrity of a structure; in fact, it's quite possible for a building or bridge to collapse even though each of its individual structural members remains intact. To ensure the stability of a structure, its designers must address two additional issues: (1) How are the pieces held together? and (2) What holds the building to the earth?

For small wood-framed structures, the most common fastener is the lowly nail. Nails hold reasonably well, provided they are kept in compression and/or shear by the loads on the building. There is, however, one way a nail will *not* hold: If you pull opposite to the direction in which it was driven, you'll pop the nail right back out. This in fact is why roofs are often lost in high winds; although most of the time a roof bears down on the rest of the structure, a high wind can reverse this direction and pull the roof members *up*. Nails alone are ineffective in preventing this type of failure.

To accommodate possible reversals in the direction of the load, particularly in regions subject to hurricanes, roof rafters and other framing timbers need to be joined by metal tie straps and brackets, and/or by drilling holes through the overlapping members and joining them with threaded bolts and nuts. In addition, triangular bracing can be quite effective in reducing lateral bending and torsional loads on a structure's joints. Such techniques are particularly important when building in coastal areas or on interior floodplains, for when a flood destroys a structure, it is usually the joints and fasteners that fail.

With larger steel and concrete structures, the design of the joints is even more critical. Bridges, elevated roadways, and tall buildings are subject to significant movement from winds and ground tremors. (For example, a long suspension span such as the Golden Gate Bridge can sway laterally as much as 4.5 meters [15 ft] on a gusty day, and the top floors of a skyscraper may move as much as 1 meter [3 ft]). Remember that such structures *need* to move a little, because it is their elasticity that allows them to adapt to changing loading conditions. At the same time, large structures must be allowed to expand and contract with temperature changes; if they don't, the resulting stresses can easily exceed the elastic limit of the steel and concrete. For instance, a 150-meter (500-ft) steel truss in Pennsylvania will expand and contract about 12 centimeters (5 in.) between a cold day in winter and a warm day in summer.

The design of fasteners for large structures is particularly challenging, because we now have two somewhat conflicting design criteria: (1) the requirement that the structural members be held together, and (2) the requirement that the joined members be allowed some degree of relative movement (usually in just one direction). Bridges and large buildings incorporate a variety of clever solutions to this problem: large encased rollers, rocker bearings, offset straps, and pin-and-hanger assemblies, to name a few.[12] In some cases, engineers have cushioned tall buildings on expansion supports that permit the entire building to rock slightly rather than bending. The mechanical details of these joints need not concern us here; my point is that there is no "perfect" method of fastening two large structural members together, because it is quite impossible to simultaneously prevent movement yet allow movement in the same joint. Compromise in design is therefore unavoidable, and in practice this usually amounts to ignoring those combinations of loading conditions that are considered to be highly unlikely to occur during the lifetime of the structure. The Northridge, California, earthquake of January 1994 triggered the collapse of sections of three major elevated freeways, and all of these failures occurred at joints or vertical supports. This does not, however, compel us to conclude that the engineering was deficient in California. Had an earthquake of this same intensity occurred in New England, the damage would have been considerably greater. (And yes, earthquakes do occur in the eastern portion of the United States, just not as often.)

The most important part of any structure is the part few people see: the foundation. A failure here can lead to the collapse of everything above. Although light one-story structures in warm climates are sometimes built on concrete slabs, usually a foundation must be much more than this. In cold climates, the foundation for even a light building must extend below the frost line, to prevent heaving and buckling when the earth expands on freezing. In Arctic regions, foundations must thermally isolate the structure from the ground, to prevent the building's interior heat from melting the permafrost; if the permafrost *does* melt, the structure will be subject to settling (probably unevenly), followed by additional movements when the supporting earth refreezes.

While it's obvious to even the most casual observer that the load-bearing capacity of soil decreases when its moisture content rises, quantitative predictions of this effect are quite difficult to make and are fraught with uncertainties. Foundations that depend on the mechanical properties of soil must therefore be built very conservatively. In coastal construction where the water table is high and the soil is predominantly sand, even relatively light homes need to be built on pilings that extend 3 to 5 meters (10–16 ft) below the ground surface. Such pilings also act as vertical beams that can resist the lateral onslaught of moderate storm surges and will stiffen the structure against high winds.

For the largest and heaviest structures, soil simply cannot be depended on to support a foundation. Most bridges and skyscrapers are therefore held up by pilings that extend through the soil and bear against bedrock, often many meters (even many stories) below the ground surface.[13] This is the best one can do, for no amount of engineering will prevent the bedrock from shifting during an earthquake or a local fault creep. Fortunately, the probability is low that the bedrock will happen to shift directly under a structure; the worst it usually does is vibrate.

If gravity were the only force acting on a structure, it would be quite sufficient to simply build the edifice so it "sits" on its foundation. In fact, around the globe, numerous historical buildings constructed in this manner have survived many centuries. Earthquakes, however, may produce lateral loads capable of knocking even the heaviest building off its foundation, while high winds and floods will transmit both lateral and lifting forces to a structure. Clearly, the engineer needs to be concerned about these possibilities. In most parts of the United States, construction codes require that buildings be rigidly connected to their foundations, in a manner that resists the lifting or base shear forces that have been identified as local disaster hazards.

Unfortunately, there is no easy way to retrofit many older structures that do not meet modern building codes. In some cases, such older noncompliant structures have proven their ability to withstand natural geophysical or meteorological events. In other cases, however, such structures may be time bombs, waiting to be set off by the next windstorm, earthquake, or flood.[14]

Dynamic Loading

Every structure must support two types of loads. *Static loads* (also sometimes referred to as "dead loads") include the weight of the structure itself plus any additional forces that act steadily on the structure (e.g., the fluid static force of the water pressing against the upstream side of a dam). *Dynamic loads* include the effects of traffic, wind, earth tremors, floodwaters, or any other rapidly variable forces that the structure could possibly experience. Clearly, it is quite possible for a structure to reliably support its static loads yet collapse catastrophically because of a dynamic load the designer didn't anticipate.

This distinction between static and dynamic loading may at first appear to be somewhat artificial: After all, from a measurement perspective, isn't a load a load? Mother Nature's answer, however, is no. Loads that are quickly applied and removed do something that static loads do not do: They induce *vibrations*. The effect is shown in Figure 3.12, where we consider three cases of a weight suspended on a spring, with a simple recording device that allows us to plot the weight's motion as time passes. In *a*, the loading is static, and the pencil's recording is just an uninteresting straight line. In *b*, we've given the weight a rapid bump (a dynamic load), and we see that it oscillates up and down for possibly quite a long time before it stops. Note that at the bottom of each oscillation, the spring is stretched quite a bit farther than it is in the static case, and that this overstretch occurs at a regular frequency even

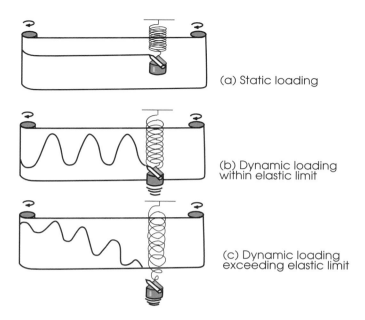

Figure 3.12. Comparison of static and dynamic loading. If the elastic limit is exceeded, the structure undergoes increasingly large deformations with each cycle, leading to eventual failure.

after the dynamic load is long gone. In *c,* we see the same dynamic situation for a spring with a lower elastic limit. Now each down cycle stresses the spring into its plastic state, from which it does not fully recover. As a consequence, each vibrational cycle stretches the spring farther and farther, long after the dynamic load that initiated the vibration has been removed.

Case *c* is the one to be avoided in structures, for it is essential that any structure return to its original configuration after the removal of a load. If it does not return, it is no longer the same structure, and all bets are off regarding the original design calculations. You may already have had the experience of being stopped in traffic on a bridge when a tractor trailer passes in the opposite direction and you felt a significant bounce, followed by a vertical oscillation. If this happens again, don't worry – the bridge is doing what it was designed to do in response to a vertical dynamic load. On the other hand, if you feel the bridge drop and *not* bounce back, get out of your car and run. Some essential part of the bridge has deformed into its plastic state, and a failure of one or more of the connecting members is almost certain to follow.

The point is this: A dynamic load will stress structural members beyond the point at which an equivalent static load will stress these same members. Placing a steady weight of, say 40 tons, on a structure demands less of the structure than rapidly applying and removing a 40-ton load. Unanticipated dynamic loads can stress structural members and their connectors beyond their elastic limits, can throw tension members into compression (and vice versa), can induce vibrations that continue for many seconds after the load is gone, and can leave the structure (if it survives) in a significantly weakened state. In fact, many buildings have been known to survive major earthquakes, only to collapse during relatively weak aftershocks.

The time required for one complete cycle of a vibration is referred to as the *period*, whereas the reciprocal, the number of vibrational cycles per second, is called the *frequency*. The period and the frequency convey the same information, and one is easily calculated from the other. For instance, if a building vibrates with a period of 0.5 seconds, its frequency is 1 cycle per 0.5 seconds, or 2 cycles per second.

The amplitude of vibration is its maximum deflection from its normal rest position. When a structural deformation remains within its elastic range, the vibrational period of the structure will be independent of the amplitude. In other words, if a building has a period of 0.5 seconds when the amplitude of vibration is 20 centimeters, the period will still be 0.5 seconds when the amplitude is just 10 centimeters. The period is a property of the structure itself and can be calculated from the weights and elastic properties of the structural members. The amplitude, on the other hand, is related to the size of the shock that initiates the vibration, which, of course, cannot be anticipated. To change a building's period, one must change the structure itself in some significant way. One may think of a building as a bell, which is rung by dynamic loading. A bell can be rung quietly or loudly, but either way it rings with the same musical pitch. In the same manner, a building will "ring" at its own natural period.

In this behavior lies an additional problem for the structural engineer: the necessity of avoiding resonance. Some dynamic loads, particularly those associated with earthquakes, have their own cyclical characteristics. If the period of an earthquake wave matches the natural period of a building, the wave will pump energy into the building's vibrations with great efficiency. This is a blueprint for disaster, for even a small earthquake under these conditions can generate a large-amplitude vibration in the building.

Mexico City, 1985

The disaster struck at 7:17 AM on September 19, 1985, killing 10,000, injuring some 50,000, and leaving 250,000 homeless, out of a population of 18 million. More than eight hundred hotels, hospitals, schools, and office buildings collapsed, most within a concentrated area of 25 square kilometers (10 mi^2). The earthquake itself was centered more than 350 kilometers (220 mi) to the west, and ground motion was not particularly severe in most parts of the city. A section near the city's center, however, stands on the site of an ancient Aztec lake, Lake Texcoco, which was drained after the Spanish conquest. The mechanical properties of the old lake bed caused it to amplify the portion of the earthquake wave that had a 2-second period, and ten cycles of this 2-second wave were transmitted to the foundations of the buildings in this region.

As always, the unreinforced masonry buildings fared poorly. Meanwhile, of the reinforced concrete structures, those under 6 stories and those over 15 stories generally survived, while those between 6 and 15 stories in height sustained heavy damage or collapsed catastrophically.

Why did this earthquake single out buildings between 6 and 15 stories in height? Because a reinforced-concrete building of this height usually has a natural vibrational period of some 1 to 2 seconds – a close match to the period of this particular dynamic load. In ten cycles of the earthquake wave, a great deal of energy was pumped into these particular structures, and they swayed back and forth like giant inverted pendulums. In fact, it was reported that some buildings continued to sway back and forth for 2 minutes, even though the earthquake disturbance itself lasted but 20 seconds. Shorter reinforced structures in the same region survived, because their natural periods were not matched to the period of the earthquake waves. Nor did the tallest building suffer any serious structural damage: The 44-story Torre Latinoamericana office building, built in the 1950s, was not vulnerable to a dynamic load with a 2-second period, because its own natural period was 3.7 seconds. The photograph in Figure 3.13 shows this skyscraper in the background, along with a tall transmission tower that survived, whereas the office building in the foreground suffered a total collapse.

The grim lesson of the 1985 Mexico City event is this: Even the strongest properly constructed buildings will be vulnerable to collapse under certain adverse conditions of dynamic loading. This presents a major dilemma to the structural engineer, for the dynamic loads that will trigger the failure of

Figure 3.13. The effects of resonance, Mexico City, 1985. Buildings between 6 and 15 stories in height sustained the heaviest damage. The 44-story building in the background was undamaged. (Photo courtesy National Geophysical Data Center.)

a given building must *not* be on Mother Nature's future agenda. Yet who can say with any confidence just what Mother Nature's program of future events may be?

Notes

1 C. W. Wright, The world's most cruel earthquake, *National Geographic*, Apr. 1909, 373–96.

2 U.S. National Oceanic and Atmospheric Administration, as cited by B. A. Bolt in *Earthquakes* (New York: Freeman, 1988), 6.

3 A number of fascinating contemporary accounts were published immediately following the 1906 earthquake, among them M. Everett, *Complete story of the San Francisco earthquake* (Chicago: Bible House, 1906), and S. Tyler, *San Francisco's great disaster* (Philadelphia: Zeigler, 1906). Numerous articles subsequently appeared in various scientific journals, and some continue to appear today: e.g., P. Segall & M. Lisowski, Surface displacements in the 1906 San Francisco and 1989 Loma Prieta earthquakes, *Science* (1990), 1241–4.

4 A portion of the Italian map is included in Wright article cited in note 1 to this chapter.

5 J. Bronowski, *The ascent of man* (Boston: Little, Brown, 1976).

6 For some fascinating examples of engineering failures that have been traced to blunders in applying reductionistic scientific principles, see M. Salvadori, *Why buildings fall down* (New York: McGraw-Hill, 1992). I also strongly recommend an earlier book by the same author, *Why buildings stand up: The strength of architecture* (New York: McGraw-Hill, 1982).

7 The *newton* is officially defined as the net force needed to accelerate a 1-kilogram mass at the rate of 1 meter per second, per second. Because this definition is somewhat difficult to apply in practice, a variety of secondary definitions, or derived standards, have been developed by the International Congress on Weights and Measures. For instance, 1 newton is also the gravitational force on a 1-kilogram mass when located at a point where the gravitational free-fall acceleration is 9.80665 meters per second, per second.

8 In common usage, the kilogram-force is often abbreviated "kg" rather than "kgf." Because the kilogram (kg) is officially defined as a unit of mass rather than force, I use the abbreviation "kgf" here when using the kilogram as a force unit. At points on the earth's surface, a 1-kilogram mass experiences a gravitational force of 1 kilogram-force, give or take 1% or 2%, depending on where it is located. A 1-kilogram mass weighs precisely 1 kilogram-force at points where the gravitational acceleration has a "standard" value of 9.80665 meters per second, per second.

9 H. R. Hitchcock, ed., *World architecture: An illustrated history* (London: Hamlyn, 1963).

10 A circle of arches does not guarantee that all of the materials remain in compression; if the circle plan is too small or a very heavy dome is being carried, the arches will be thrust outward, and tension failures can arise in the upper courses of masonry. Masonry domes on some old buildings have had to be banded with steel to prevent complete failure from this effect.

11 The first comprehensive account of this disaster was F. Connelly & G. C. Jenks, *Official history of the Johnstown Flood* (Pittsburg: Journalist, 1889). A

more accessible source is D. McCullough, *The Johnstown Flood* (New York: Simon & Schuster [Touchstone], 1968). For a shorter account of this disaster, I refer the reader to William H. Shank, *Great floods of Pennsylvania: A two-century history* (York, Pa.: American Canal and Transportation Center, 1972).

12 The failure of a pin–and–hanger assembly on a section of an Interstate 95 bridge in Connecticut in 1983 led to the collapse of one span of the bridge and three deaths. For a description of this failure and other examples of problems with structural joints, see H. Petroski, *To engineer is human: The role of failure in successful design* (New York: Vintage, 1992).

13 One of the piers of James Eads's bridge over the Mississippi at St. Louis extends 41.5 meters (136 ft) below the level of the river. The base of the New York tower of the Brooklyn Bridge is 24 meters (78.5 ft) below the water.

14 For discussions of this issue, see K. Matso, Lessons from Kobe, *Civil Engineering,* Apr. 1995, 42–7, and G. Zorpette, Bracing for the next big one, *Scientific American,* Apr. 1995, 14–16.

4

Death and Life

The Killer Lakes of Cameroun

Two strange and silent disasters struck villages in the Republic of Cameroun in West Africa in 1984 and 1986. Scientists had barely unraveled the mysteries of the first, which killed 37, when a nearly identical catastrophe claimed more than 1,700 additional human lives. The victims did not die from trauma, extremes of temperature, disease, or starvation. They died, many of them under the open sky, from lack of oxygen.

The first disaster occurred on August 16, 1984. At 5:30 AM, the police department in the town of Foumbot received reports that people were collapsing on a road at nearby Lake Monoun. When several officials and the local physician reached the site an hour later, they saw bodies scattered along the road, and a white, ground-hugging cloud drifting toward them with the wind off the lake. As the cloud approached, the investigators began to feel nauseated, dizzy, and weak in the legs, and they quickly (and wisely) retreated until the strange haze dissipated. Later they reported that when they finally did examine the victims, many had skin lesions or blisters, foam at the nose or mouth, distended stomachs, and evidence of incontinency. In the immediate area they also found numerous dead rats, bats, snakes, and at least one dead cat.[1] Unfortunately, no autopsies were performed on any of the victims, and no tissue samples were taken.

Because some sort of geophysical event seemed to have taken place, several groups of geoscientists visited the region in the following months and attempted to integrate the local eyewitness accounts with after the fact chemical analyses of water and sediment samples taken from the lake. Soon it became apparent that no single scientific hypothesis could possibly be consistent with all of the data and eyewitness interviews. In particular, any gas that could conceivably have arisen from the lake would not have caused the whole range of reported physiological damage to the victims; moreover, had acids been involved, clothing and vegetation would also have borne telltale

93

evidence long after the event. Given that the several victims who survived did not experience foaming or skin lesions, we have to suspect that the reports of the officials were inaccurate on some details. Other eyewitness accounts gathered by the scientists, however, seem to be quite relevant: At 11:30 PM on the night before the disaster, numerous residents of two nearby villages heard a distinct rumbling sound from the vicinity of the lake.

Viewed from the surface, Lake Monoun is a relatively small body some 1.5 kilometers in length, with a width that varies between about 200 and 700 meters. Cradled in a steep-walled volcanic crater, the lake is fairly deep relative to its surface dimensions, 96 meters (315 ft) at its deepest point (considerably deeper than Lake Erie, for instance). The rumbling noise seems to have been generated by a landslide that dropped a large volume of rock and dirt into the deepest part of the lake. Not only would such a landslide have created a great surface wave (and in fact there was evidence in the vegetation of a recent 5-meter wave), but beyond this, such an impact would certainly have churned the lake's deepest water to the surface.

Occurring in the middle of the night, neither the landslide nor its resulting wave claimed any victims directly. But unfortunately, Lake Monoun's deep water was saturated with dissolved carbon dioxide, which had entered over many years through cold volcanic vents in the crater floor. When this deep, pressurized water was rolled to the surface, the dissolved carbon dioxide immediately fizzed out, just as it does when you open a can of carbonated beverage after shaking it.

Of course, carbon dioxide itself isn't poisonous; the earth's atmosphere always contains a small amount of this gas, which is essential to plant life. But because carbon dioxide is some 40% denser than air, it flows into low places and displaces ordinary atmospheric air.[2] On that tragic morning in 1984, the gas cloud that burst from the the lake slithered along a river valley to the east and suffocated most of its victims as they approached a low bridge around daybreak.

The Lake Monoun disaster was a relatively obscure event that attracted virtually no media attention when it happened. The phenomenon did attract a number of scientists, but scientific studies take time. Before any findings or hypotheses were reported in the scientific journals, a second and more serious disaster struck at the site of another deep crater lake in Cameroun: Lake Nyos.

> On 21 August [1986] at about 2130 [9:30 PM] a series of rumbling
> sounds lasting perhaps 15 to 20 seconds caused people in the immedi-

ate area of the lake to come out of their homes. One observer reported hearing a bubbling sound, and after walking to a vantage point he saw a white cloud rise from the lake and a large water surge. Many people smelled the odor of rotten eggs or gunpowder, experienced a warm sensation, and rapidly lost consciousness. Survivors of the incident, who awakened from 6 to 36 hours later, felt weak and confused. Many found that their oil lamps had gone out, although they still contained oil, and that their animals and family members were dead. The bird, insect, and small mammal populations in the area were not seen for at least 48 hours after the event. The plant life was essentially unaffected.

. . . Vegetation damage showed that a water surge had washed up the southern shore to a height of ~25 m. A water surge 6 m high had flowed over the spillway at the northern end of the lake, and a fountain of water or froth had splashed over an 80-m-high rock promontory on the southwestern shore.[3]

This second gas cloud spread as far as 10 kilometers (6 mi) from Lake Nyos and killed 1,700 people and approximately 3,000 cattle. This time, scientists began arriving within a few days and were able to establish beyond any reasonable doubt that the victims had died from carbon dioxide asphyxiation. Although sulfides may have been present in the gas cloud (possibly accounting for the odor), they were not present in sufficient concentrations to cause death. The skin lesions found on some victims were attributed to preexisting tropical diseases rather than heat or chemical burns. No direct volcanic activity or seismic event could be blamed for the sudden release of the gas. A fresh landslide scar was found on the western cliffs facing the lake, suggesting that the event was probably precipitated in the same manner as at Lake Monoun. Lake Nyos measures 1,925 meters in length, 1,180 meters at its maximum width, and 208 meters in depth, making it larger and deeper than Lake Monoun. This adequately explains the larger volume of emitted gas, and in turn the larger death toll.

Today, the deep waters of both lakes continue to hold large quantities of dissolved carbon dioxide, and the threat of a repeat occurrence is very real. There does, however, exist a plausible strategy for avoiding another disaster of this type: to continually pump water from the depths to the lake's surface, where the carbon dioxide will effervesce continuously but at a rate low enough to prevent a wholesale displacement of the atmospheric oxygen in the surrounding region. The engineering principles are quite simple. Unfortunately, the capital for such a project is not currently available. (The per capita Gross Domestic Product of Cameroun is less than 5% that of the United States.) The best that can be done at present is to monitor Cameroun's crater lakes with increased vigilance, in the hope that timely evacuation warnings can be given if any of these waters should ever again regurgitate a large burst of suffocating carbon dioxide.

Population Distributions and Disasters

In humankind's early years, when survival depended on hunting and gathering and social groups were small and geographically dispersed, major natural disasters would have been the least of anyone's worries. Before farms and cities, there was little to tie any group to a particular geographical area. If a volcano began to rumble, a tribe could easily move on with no sense of loss; if an earthquake struck, there were no heavy buildings to tumble onto sleeping children. Epidemics were unlikely, because contagious diseases require a threshold host population to sustain the survival of the microbes. Tsunamis and floods might claim casualties, but never in great numbers, simply because in a hunter-gatherer society significant numbers of humans never congregate in one place at one time. A necessary prerequisite to natural disaster is a large number of humans living semipermanently in one place.

Although our planet is usually generous about sustaining human life, the forces of nature do occasionally transform regions of our environment into death traps. When one of these environmental anomalies coincides with a dense pocket of human population, we have a natural disaster. Blizzards in Antarctica, waterspouts in the Atlantic Ocean, and volcanoes in Siberia may be dramatic events, but they are seldom disasters. A disaster occurs when a blizzard strikes a large city, a tornado hits a mobile home park, or a volcano explodes at sea level and sends tsunamis racing toward inhabited shorelines. One cannot talk about the risk of disaster without considering where people have chosen to live and how densely they have packed themselves together.

The current population of the world stands at around 5,700,000,000 people (5.7 billion), nearly one-fifth of whom live in China. The second most populated country is India, followed by the United States, then Indonesia. Population figures for nations, however, don't give us a true picture of global population distribution, because different countries are different sizes. A somewhat better measure is the population per unit of land area. In Table 4.1, I've ranked some representative countries in order of their average pop-

Table 4.1. *Population and average population densities of selected nations.*

Nation	1993 population (millions)	Population per km^2
Bangladesh	119	824
Taiwan	20.9	581
South Korea	44.2	448
Netherlands	15.1	370
Japan	124.5	321
India	886.4	270
Great Britain	57.8	237
Israel	4.8	234
Philippines	64.1	224
Italy	57.9	192
North Korea	22.3	184
Poland	38.4	122
China	1,170	122
Indonesia	195	101
Egypt	56.4	56
Mexico	92.4	47
Ethiopia	51.1	45
United States	257	27
Cameroun	12.6	27
Russia	149.5	8.5
Canada	27.4	2.7
Australia	17.6	2.3
World	5,700	36.7
World, excluding Antarctica	5,700	40.4

Note: 1 square kilometer = 0.386 square miles.

ulation densities, in people per square kilometer. By this measure, the most densely packed country in the world is Bangladesh; on average it is 8 times as crowded as China. In fact, China slips far from the top of the list when we rank countries by population density rather than by total population.

Table 4.1 show that the most densely populated regions of the world are in southern and eastern Asia, the western Pacific islands, and western Europe. In contrast, Africa, the Americas, and Australia have much lower average population densities. It is nevertheless a sobering fact that we humans have been so successful in multiplying our numbers that today the average population density of the world as a whole (excluding Antarctica) is more than 40 people per square kilometer (100 per mi^2). This suggests that there remain very few places on our planet where nature can go on a rampage without affecting a large number of our species.

Still, it's important to remember that even within a single nation the average population density varies significantly from place to place. Obviously, the greatest population densities are found in cities (which is what makes them cities). The most densely populated city in the world is Hong Kong, with its average of 95,535 inhabitants per square kilometer (247,500 per mi^2) nearly 22 times the density of New York City and more than 27 times as crowded as Los Angeles. Next to Hong Kong, the world's most densely populated cities, in order, are: Lagos, Nigeria; Dhaka, Bangladesh; Jakarta, Indonesia; Bombay, India; Ho Chi Minh City, Vietnam; Ahnadabad, India; and Shenyang, China. In places like these, enormous numbers of lives are dependent on the continued integrity of the physical infrastructure. Should Mother Nature knock out the power, interrupt the water system, block the transportation of food, shake buildings to the ground, or start a few fires, the impact on many, many inhabitants will be catastrophic.

People who live in regions of high population density are extremely vulnerable to breakdowns in their web of dependency on each other. In a sparsely settled farming community, a severe earthquake may render considerable damage to structures, but survivors will have no problems finding food and water to sustain themselves, and there won't be any major obstacles to making temporary repairs to their homes and barns. Contrast this with the effect of a similar event on a high-rise apartment dweller in a large city. Food may become unavailable because highways and bridges have been damaged; water service may be interrupted (or water supplies contaminated); and structural repairs to the reinforced-concrete building will be far beyond the capabilities of the apartment residents themselves. Meanwhile, the individual survivor must compete with many other survivors for the diminished life-supporting resources, including temporary housing. To live in a region of high population density is always to sacrifice self-sufficiency. Under such circumstances, most survivors of a disaster have little choice other than to wait for government-organized relief efforts.

In a prosperous nation (one, say, with a high GDP per capita), disaster relief is usually fairly quick and effective, in spite of the inevitable bureaucratic complications. Most survivors are soon fed, clothed, and provided with medical care and temporary shelter. Seldom in a developed country is a disaster followed by an outbreak of cholera or any of the other diseases associated with poor sanitation. But in poorer nations, where the effect of a disaster is likely to be aggravated by lax building codes and an overstressed infrastructure, relief efforts are often minimal and/or otherwise ineffective. Thus do economic factors contribute considerably to the human suffering and ultimate death toll of disasters in the underdeveloped nations of the world.

The Floods of Bangladesh

The nation of Bangladesh, contiguous to India's northwestern border, has a population of 119 million people in an area smaller than the state of Wisconsin (giving it, as noted earlier, the highest average population density of any country in the world). About 80% of the region is a broad, flat plain perched just slightly higher than sea level, divided into thousands of islands by the Ganges and Brahmaputra Rivers, their deltas, and their hundreds of tributary rivers and streams.[4] The climate is one of the rainiest in the world, and the rivers (which combine the waters from the interior of India on the west and from the Himalaya Mountains on the north) flood quite regularly during the monsoon season. If the major rivers happen to flood simultaneously, Bangladesh has a disaster. If the rivers flood while tides in the Bay of Bengal are high, the disaster is aggravated. Should a tropical storm coincide with these events, the human impact can be enormous.

A tropical storm, because of its low barometric pressure, raises the tides to abnormal heights. Since free-running water can only flow downhill, such extreme tidal bulges block the egress of water from the river deltas. The resulting backup is much more serious than a run-of-the-mill flood; it is a flood with violent wind-driven waves that sweep far inland from the now-submerged coastline. Even structures built on stilts are torn from their foundations. In a place like Bangladesh, where there is precious little high ground for evacuees to escape to, this combination of conditions can, and too often does, lead to a major catastrophe.

In the last thirty-five years, Bangladesh has suffered at least seven major natural disasters that fit the basic scenario just described:[5]

May 28–9	1963	22,000 deaths
May 11–12	1965	17,000 deaths
June 1–2	1965	30,000 deaths

December 15	1965	10,000 deaths
November 13	1970	300,000 deaths
May 25	1985	10,000 deaths
April 30	1991	131,000 deaths

As is characteristic of disasters, these death tolls reveal only a small part of the human tragedy. Of the present inhabitants of Bangladesh, over half have been rendered homeless by the forces of nature at least once in their lifetime, and numerous families have had this happen multiple times. For instance, in the years 1988 and 1989 alone, when "only" 4,000 died due to the flooding, as many as 30 million were left homeless; in the great flood of 1991, when 131,000 humans perished, inhabitants also lost more than 500,000 livestock, and 10 million lost their homes. Moreover, Bangladesh has only 1 physician per 5,500 people (compared to 1 per 404 in the United States), and physicians themselves are by no means immune from being rendered victims of floods. As a result, each of this country's major floods has been immediately followed by outbreaks of untreated disease that have claimed many additional lives.

Of the major floods worldwide during the twentieth century that have claimed 10,000 or more lives, all but three or four have occurred in what is today Bangladesh. This historical pattern suggests that some engineering response ought to be considered, perhaps along the lines of the dikes and sea gates the Dutch have built to protect their own low country against the North Sea. Unfortunately, Bangladesh is much too poor to maintain a quality dike system. The Netherlands has a population of 15 million, or about 13% of the population of Bangladesh, yet its Gross Domestic Product is $175 billion per year, more than 11 times as great as Bangladesh's $15.6 billion. The available capital in Bangladesh is grossly insufficient to fund a major (and incredibly difficult) engineering response to that country's recurring disasters. Another great flood is sure to strike, and thousands more will surely die.

This desperate state of affairs may sound like a death knell for the nation, for if tens of thousands die in major flooding every few years, won't the population eventually shrink to virtually zero? The answer is no. In fact, in recent years the population of Bangladesh has actually been increasing at a rate of 2.3% per year. Each year, in other words, the population grows by about 2.3% of 119 million, or about 3 million. Subtract a few tens of thousands killed in a tropical storm, and it hardly affects the growth trend at all.

Yet as the population continues to grow, increasing numbers of people are vulnerable to the next tropical storm, and the prospects for protecting them are increasingly remote. Bangladesh has the dubious distinction of ranking first among the world's nations in potential for human misery due to future natural disasters.

The Natural Law of Growth

The arithmetic of geometrical growth can lead to astounding and frightening conclusions. Consider the following story puzzles:[6]

Story 1. A king, wanting to reward a faithful servant, asks the man to state his wish. The servant, noticing the king's chessboard, responds that he would like 1 grain of rice for the first square on the board, 2 grains for the second square, 4 grains for the third, and so on, doubling the number of rice grains for each square up to the sixty-fourth square. The king quickly agrees to so humble a request, and directs one of his ministers to go to the granaries to count out the rice. Several hours later the minister returns to the king in a panic. It turns out there isn't enough rice in the kingdom, or in the entire world for that matter, to grant the servant's wish! For the sixty-fourth square alone, the servant is entitled to 9,223,372,000,000,000,000 grains of rice, which would weigh in at around 7 billion tons!

Story 2. An ambitious but inexperienced reporter begs an editor for a job at a major daily. When asked what she feels would be a fair salary, in view of her lack of experience, the writer responds that she'll accept a penny for the first day's work, two cents for the second day, four cents for the third, and so on, doubling her salary each day for the first four weeks, or twenty-eight days. The editor quickly agrees to so modest a request and sends the directive to the payroll department. As a result, by the end of the month the young reporter has earned enough to retire. Here's how her salary develops:

Day	Daily salary	Weekly totals
1	$.01	
7	.64	1st week total = $1.27
14	81.92	2nd week total = $162.56
21	5,242.88	3rd week total = $10,403.84
28	335,544.32	4th week total = $665,845.76

Story 3. A water plant germinates on the surface of a pond, and each day it doubles in size (that is, each cell divides once each day). In thirty days, this plant will completely cover the pond. If the pond's owner decided to eradicate the plant when it covered just half the pond, on what day would this occur? No, there is no trick here; obviously, the plant covers half the pond on the twenty-ninth day. The moral, however, is this: The imminent crisis of the pond's disappearance is not apparent until there is precious little time left to act.

As long as sufficient resources are available to support life, human populations will grow geometrically in time. In other words, the number of people added to a population depends on how many people are already alive: The larger the population, the more mating takes place and the greater the number of births that drive future population growth. If, however, growth continues unabated in this manner, eventually a population is sure to outstrip the regional resources needed to support it. This is what seems to be happening in places like Bangladesh, a country that may already be well into its metaphorical twenty-ninth day.

Our interest in population growth applies to microbes as well as to people. A man may be infected with a growing colony of microorganisms for quite some time with no symptoms; then he wakes up one morning and suddenly feels like he's been run over by a freight train. The natural law of growth tells us that there can be a drastic difference between the twenty-ninth day and the thirtieth day.

But do populations really double all that quickly? "Quickly," as we have seen, is a relative term. The graph in Figure 4.1 shows that if we look at the extended picture, the human population of our planet is indeed growing dramatically, with most of the growth having taken place within just the last few generations.

A useful concept here is that of "doubling time." If a population is consistently growing, it will eventually double. How soon it will double depends on the net growth rate, which is most conveniently expressed as a percent:

$$\text{Net growth rate (\%)} = \text{birth rate (\%)} \quad \text{death rate (\%)}$$
$$+ \text{ net immigration rate (\%)}$$

For instance, if an island nation has a birth rate of 10% per year, a death rate of 2% per year, and a net immigration rate of 1% per year (i.e., people are

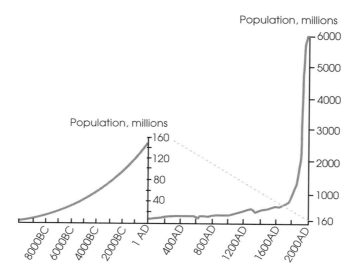

Figure 4.1. The growth of world population. The current growth rate is about 1.8% per year, which results in a doubling every thirty-nine years.

leaving), then the net growth rate is 7%. This may not sound like a very high rate, but in fact it is. Watch what happens if a population of 100,000 increases by 7% each year:

Year	Population
0	100,000
1	107,000
2	114,490
3	122,504
4	131,080
5	140,255
6	150,073
7	160,578
8	171,819
9	183,846
10	196,715
11	210,485

This arithmetic shows that a growth rate of "just" 7% per year leads to a doubling of the population in about 10 years. Similar arithmetic will establish that at a growth rate of 3.5% per year, the population will double in 20 years. In the case of Bangladesh, with its 2.3% growth rate, the population doubling time is about 30 years.

There is a simple way to arrive at the doubling time without going through a long string of arithmetic. The formula is this:

$$\text{Doubling time} = \frac{70}{\text{growth rate in \%}}.$$

For instance, a growth rate of 2% per year doubles the population in 35 years, and with a growth rate of just 0.5% per year, the population will still double in around 140 years.[7] Notice that the only growth rate that does not lead to an eventual doubling is 0% (or less).

At present, the world population is around 5.7 billion, and the growth rate stands at about 1.8%. If nothing is done to reduce the growth rate, we can expect the world population to grow to 11.4 billion in about 39 years. Take a look at a map, and imagine doubling the number of towns and cities, or doubling all their sizes, in just 39 years. Viewing this another way, how many of us can honestly imagine adding 10 cities the size of Los Angeles to the globe *each year?* Yet this is precisely what we're doing on our planet already. In future years, given current trends, the prospective scenario gets worse: We will add 20 equivalents of the Los Angeles population each year, then 40, then 80, and so on.

The unfortunate consequence of rapid population growth is that cities with inadequate infrastructures have become increasingly common around the world, and a rapidly growing number of humans are finding themselves dependent on overstressed city infrastructures to provide their basic needs. Many of today's cities already lack adequate water systems, sewer systems, medical services, and food and fuel distribution networks to serve the daily needs of their current populations. Problems like power outages and transportation tie-ups are not only symptomatic of the population pressures but, under even the best circumstances, they create widespread inconvenience if not outright hardship. Now add one of Mother Nature's hiccups to this precarious state of affairs. The conclusion is inescapable that as city populations continue to grow, they will become increasingly vulnerable to natural disasters. Perhaps this will ultimately be Mother Nature's method for getting us humans back into equilibrium with our planet.

One thing is certain: Current population growth trends cannot continue indefinitely. If they did, it's easy to predict that the world's current population would double to about 11 billion by the year 2033. By the year 2072, our children and their offspring would see another doubling, to 22 billion. Around the year 2160, the entire land area of the world, including deserts and glaciers, would be as densely populated as today's country of Bangladesh. In 570 years, our standing room would be reduced to an average of just 1 square meter (11 ft^2) per person, and in order to lie down to sleep we'd need to do it in shifts. Of course the scenario gets ridiculous long before this point, because even the most inflated estimates of the carrying capacity of Planet Earth suggest an absolute upper limit of somewhere between 10 and 15 billion people.[8] We are, in other words, already in the twenty-ninth day.

The Easter Island Disaster

Only in the last few years have scientists pieced together a credible account of the disaster that befell Easter Isand, an isolated little South Pacific island. This was a gradual event, as natural disasters go, spanning several generations and perhaps even a few centuries. The grim result was the irreversible destruction of an entire culture and the ecosystem that had supported it, and the deaths of many thousands of humans. Forests and food crops disappeared; for lack of timber, ship construction ended (and along with it, the harvest of sea food); social organization disintegrated; and people took refuge in caves, because there were no longer any building materials to provide shelter. A society that had once cooperated in the construction and transportation of immense stone monuments reverted to tribal warfare and cannibalism, and ultimately even the oral traditions of the earlier civilization were lost.

There is no inhabited place on earth more isolated than Easter Island. The natives living there in the 1700s did not even have a name for their homeland, for the simple reason that they were unaware that any other place existed. They would have had to travel 3,200 kilometers (2,000 mi) to the east to reach South America and about 2,200 kilometers (1,400 mi) to the west to reach the first islands in the Tuamotu Archipelago. Yet their only boats were so primitive and leaky that they became unseaworthy after only a few hours in the water. As far as the eighteenth-century natives knew, their 160 square kilometer island and their view to the horizon constituted the entire world.

When the Dutch explorer Jacob Roggeveen first landed on the island, on Easter Sunday, 1722, and when, fifty years later, the British arrived under Captain Cook, approximately two hundred giant stone statues with similar stylized human features stood as lonely sentinels facing the sea (Fig. 4.2). Some of these monuments were as tall as 10 meters (33 ft) and weighed as much as 82 tons. Many stood in rows on huge stone-faced platforms up to 150 meters (500 ft) in length and 3 meters (10 ft) high. In the quarries, which were up to 10 kilometers distant from the erected monuments, at least seven hundred additional statues were found abandoned in various stages of completion. Some of these were as tall as 20 meters (65 ft) and weighed up to 270 tons. How could such a primitive people, who numbered only around 2,000 (as reported by the first explorers) and who spent most of their time scrounging for food and fighting each other, have carved, transported, and erected such grand monuments?

Figure 4.2. The *moai,* megalithic sculptures of Easter Island, are remnants of a civilization that was destroyed by the environmental consequences of its own overpopulation.

Clearly, they couldn't have. The main quarries are at the northeastern end of the island, the monuments at the other end. Red stone used for crowns on some of the statues came from an inland quarry in the southwest, while the stone carving tools came from the northwest. The best farmland was in the south and east, the best fishing on the northern and western coasts. Obviously, there had once been an effective central political organization that governed the island's whole population. That population probably numbered close to 20,000 at its peak,[9] and all of these people had at some remote time worked together rather than at cross-purposes.

So what happened? An answer consistent with Occam's Razor can be pieced together by combining the results of archaeological excavations, pollen counts in the excavated layers of soil, radiocarbon dating, and DNA analyses of human remains. Combining all of the current evidence gives us the following general story:[10]

Sometime between A.D. 400 and A.D. 700 (no doubt encouraged by their past successes in always having found another island to settle), the Polynesian seafarers who had colonized Fiji, Samoa, and Tahiti continued to push eastward into uncharted waters. This time, however, there just wasn't another island within a reasonable distance. Probably most of the embarking vessels never did make another landfall, and their occupants perished in the unexpected vast emptiness of the southeastern Pacific Ocean. A few of these Polynesian adventurers did, however, chance to stumble upon Easter Island, and they brought ashore their customary cargoes of banana plants, taro, sweet potatoes, sugarcane, mulberry plants, and chickens, along with the ubiquitous rat.

At that time, most of Easter Island was covered by a lush subtropical forest. Numerous hauhau trees provided the raw material to make rope, and toromiro trees supplied a dense firewood. The most common tree seems to have been a now-extinct palm that grew to a height of 25 meters (82 ft) and a diameter of 2 meters (6 ft), its trunk ideal for boatbuilding, its nuts edible, and its sap sweet and nutritious. The island was a breeding site for at least twenty-five nesting species of birds, and there is evidence that seals once lived here in sizable colonies. At 27 degrees south latitude, Easter Island is just outside the tropics, and the local waters are a bit too cool to support the coral reefs that support great numbers of tropical fish. But even here Easter's Polynesian colonists were fortunate, for large numbers of porpoises could be found in the deep waters just a few kilometers out to sea from the island. The new immigrants were obviously a people who knew how to build seaworthy boats that allowed them to go out and harpoon the animals. Initially, then, Easter Island was a bountiful place to establish a human settlement.

By A.D. 800 Easter's deforestation was well underway. Still, until at least 1300 food remained plentiful. Garbage dumps from this period tell us that the meat diet consisted of about one-third porpoise, one-fourth fish, and the remainder seabirds, land birds, shellfish, rats, domesticated chickens, and possibly an occasional seal, all in addition to an ample vegetable diet. The population multiplied and provided an abundant pool of labor for an ambitious public works project. This project, the transportation and erection of the giant stone statues, in turn accelerated the depletion of the forests that the whole society was so dependent upon. Wood was now needed for sleds and levers in addition to boats and dwellings, and the rope required to haul the statues and raise them into position accelerated the extinction of the hauhau tree.

We don't know when the first statues were carved and raised, but most seem to date from around 1200 to 1500, with the general abandonment of the quarries closely following the year 1500. Pollen records tell us that the last large palm trees disappeared soon after 1400. As the forest disappeared, soil erosion increased, springs and streams began drying up, and crop production declined. Porpoise bones are no longer found in garbage heaps after around 1500; with no large trees remaining, it was no longer possible to build boats that would take the natives out into the deeper waters where they could harvest these sea creatures. The islanders turned to wild birds as a source of food and soon drove them to extinction. This led to an increasing dependence on domesticated chickens and, eventually, to the consumption of human flesh. Surviving statuettes from the 1600s depict people with sunken cheeks and visible ribs, suggesting that many indeed were starving.

When people begin eating each other's relatives, they are much less likely to cooperate in solving mutual problems. In the 1600s and 1700s, so many spearpoints and stone daggers were chiseled that today, centuries later, they are still not difficult to find in the soil. The implication of these artifacts is quite different from the discovery of arrowheads crafted by Native Americans who used them to hunt bison and other animals, because on Easter Island the only animal left in any significant number was *homo sapiens*. Humans became both hunters and hunted, on an everyday basis. Charcoal residues in the soil reveal mass burnings of human constructions. Families began living in caves, and the political system degenerated into anarchy. By 1700, the population had crashed from around 20,000 to about 2,000.

Strife continued even after the first contact with Europeans in 1722. Although two hundred of the huge statues were still standing in 1770, by 1864 all of them had been toppled by rival warring clans. Only after ships began to stop and provide supplies on a fairly regular basis did the islanders settle into their current state of tranquillity.

Take a few people, put them on a remote but bountiful island where they have no practical way of escaping, let them multiply for fifty generations, and their children eventually outstrip the carrying capacity of their homeland, then turn on each other and destroy all that has been accomplished in generations past. No one sees it happening until it's too late, the twenty-ninth day. Extinct animals and trees do not come back, and the civilization that depended on them is gone forever. So it was with Easter Island.

Take a few people, put them on a remote but bountiful *planet* from which they have no practical way of escaping, let them multiply for a few thousand generations, and eventually their children will likewise outstrip the finite carrying capacity of their planetary homeland. Will we also turn on each other in our mad scramble for survival and destroy all that has been accomplished in generations past? Some argue that it may already be the twenty-ninth day. The Easter Island disaster may indeed be about to be repeated on a global scale, driven by the unbridled multiplication of the world's human population.

Evolution and Natural Selection

The paradigm that new natural forms arise continuously from earlier forms is found in all of the sciences. On the scale of the Cosmos, we see evidence of the evolution of galaxies and the stars within them. On a planetary scale, we study mountains and canyons and find evidence of geological evolution. Societies, cultures, and economic systems are also seen to evolve; new forms arising from preexisting forms. Evolution refers to the observation – the "fact," if you will – that things change continuously from one form to another, and that there is an underlying statistical pattern to such changes.

Biological evolution is a phenomenon, not a theory. The fossil record makes it quite clear that many life forms of past ages no longer exist, and conversely that most of today's life forms did not exist in the remote past.

Moreover, modern biochemical analysis makes it possible to confirm the continuity of life, establishing, for instance, that humans are distant cousins of both olive trees and amoebas.

Regardless of the life form, the conveyor of genetic information from one generation to the next is the DNA molecule, whose structure was first deciphered in 1953.[11] This giant molecule contains many millions of atoms of nitrogen, phosphorus, carbon, hydrogen, and oxygen, but all of these atoms are joined together in just four recurring configurations, which may be visualized as the rungs of a long helical ladder that twists clockwise in the same sense as a standard right-handed screw thread does. Closely related organisms are found to have similar sequencing of these ladder rungs (the so-called base pairs). The more distantly related the organisms, the more differences one finds in the sequencing of the base pairs in their DNA.

DNA has the ability to replicate; that is, to create an accurate chemical copy of itself, if the surrounding fluid of the cell contains the right ingredients. During replication, the molecular pairs forming the rungs of the DNA ladder separate and grab a new partner from the surrounding chemical soup: each adenine grabbing a thymine molecule, and each guanine latching onto a cytosine. The result is a pair of DNA molecules with identical base-pair sequences, that is, the same genetic information as the original DNA molecule. These daughter DNA molecules both twist in the same direction: clockwise, as one looks up the ladder.

From a purely chemical perspective, there is no reason for DNA to be twisted clockwise rather than counterclockwise. A counterclockwise DNA molecule would just as effectively undergo all of the same chemical interactions with its environment and would replicate to form new counterclockwise DNA molecules. And yet, in the many thousands of samples of genetic material that have been analyzed from thousands of species of living organisms, the DNA is always found to twist clockwise, never counterclockwise. This is also the case with those samples of DNA that have been extracted from the preserved remains of extinct life forms.

Now this is very curious, for if there are no chemical or physical reasons for DNA to twist clockwise rather than counterclockwise, then why is only one form of this molecule found in living organisms? The replication process explains why a mother and her son will have their DNA twisting in the same direction, but why should organisms as diverse as banana trees and lobsters also have only clockwise DNA? The probability of this happening by pure chance is much less than having everyone in the world simultaneously flip a coin and having them all land heads up. There is only one plausible explanation: Mother Nature's very first DNA molecule just happened to spiral clockwise, and all current DNA molecules are daughters. All life, in other words, has the same origin.

This biochemical evidence complements the earlier fossil evidence that new life forms evolve, usually very slowly, over many, many generations. Evolution is as real a phenomenon as lightning or earthquakes, and the observational basis is every bit as solid. Evolution just happens to grind along so slowly that we don't usually notice new life forms arising within a human lifetime. Notice I say "usually," for we will see later that some disease-causing microorganisms do indeed evolve into new forms relatively quickly on even human time scales.

At this point, however, we need to make a careful distinction: The *phenomenon* of evolution is no more a *theory* of evolution than the phenomenon of lightning is a theory of lightning. A theory is a great cognitive leap beyond the level of the observations themselves; a theory seeks to explain the mechanisms that drive the phenomenon. The prevailing scientific theory of evolution, the *theory of natural selection,* traces its roots to the writings of the naturalists Charles Darwin (1859) and Alfred R. Wallace (1876). In the days of Darwin and Wallace, it was difficult to formulate a theory of evolution in language that was empirically falsifiable, and progress was slow. Many scholars argued that the idea of natural selection did not qualify as a theory at all, since it did not predict anything that could be tested objectively. There was some validity to this point of view, and until quite recently the primary reason for the popularity of the theory of natural selection was the lack of any viable alternative theory. Today, however, advances in genetics and biochemistry have made it possible to devise experimental tests of some of the consequences of natural selection, and the theory is on much firmer ground.

The theory of natural selection does not deal with the fate of specific individuals; it deals only with the probabilities of survival of large populations of organisms. The theoretical mechanism is summarized in Table 4.2. The first requirement here is that there must be some degree of genetic variation within any cohort; that is, all cousins cannot have identical traits. We now know that such variation always occurs to some extent, even in the lowest of life forms, through occasional mistakes (mutations) that enter the DNA molecule during its replication. Thus, just by chance, some individuals will be better adapted to their environment than their cohorts (having slightly keener vision, for instance, or a slightly better sense of balance). Meanwhile, just by chance, other individuals born at the same time will be at a slight dis-

Table 4.2. The theoretical process of natural selection. *Any species can be expected to acquire new biological traits over a series of generations as a result of the following naturally occurring sequence:*

1. Genetic variation among members of the species.

2. An environment that kills some individuals before they reproduce, while favoring the survival to maturity of other individuals whose genetic traits are better adapted to the environmental pressures.

3. Reproduction by the survivors at a rate that exceeds the parental replacement rate.

4. The transmission of genetic information from the parents to the off-spring during reproduction.

5. The ultimate physiological decline and death of the parents, which removes them from the cycle of competition and sifts their temporally acquired traits from the long-term genetically determined traits of the species.

advantage (having weaker claws, for instance, or being incapable of jumping as far).

The second requirement is that only a portion of any cohort survives to reproductive maturity; again, this usually seems to happen. Because random accidents inevitably claim some individuals, survival of the "fittest" is by no means guaranteed. What is statistically likely, however, is that a higher proportion of the better-adapted individuals will succeed in evading predators, enduring an extreme winter, fighting off an infection, and so on. This introduces a statistical bias in which the longer-lived individuals are more likely to have those traits that favor the survival of the organism.

Next, the organism must reproduce in a manner that transmits most of its own genetically inherited traits to its offspring. Moreover, if the life form is to evolve, it must reproduce at a rate that exceeds its own replacement rate (that is, each parent must average more than one offspring). If there is no excessive reproduction, the natural selection process will result in a continually dwindling population and eventual extinction of the species.

Finally, natural selection requires the decline and death of the individual. If a species is to evolve, its individual organisms *must* have a finite lifetime. If the old didn't decline physically and die, they would continue to compete for survival in an ever-burgeoning population of offspring. In any life form capable of even rudimentary learning, the best-adapted individuals are sta-tistically the older (they've had more time to learn), up until the age where this advantage begins to be undermined by physiological decline. In the absence of aging, evolution would grind to a halt, because the new genetic traits would have little chance of replacing the learned traits of the still-alive earlier generations. When the environment changes (and it always does, eventually), any slowly aging and slowly reproducing species has little capac-ity for adaptation, and it disappears into extinction. To be replaced by what? A species that followed a different evolutionary branch to arrive at a form where its individuals are genetically programmed to die soon after they've reproduced and nurtured their young to maturity. In the theory of natural selection, it is aging and death that drives the evolutionary process.[12]

Does a species *need* to evolve in order to survive? Absolutely. In ages past, our planet was a very different place from today. The seas were not as salty, the early atmosphere had much less oxygen, and (before the more recent cycle of ice ages began) the average temperature was higher than today's. A species that thrives in one environment is always at risk when the environ-ment changes. Life forms that do not change, or that change more slowly than their environment changes, are statistically more likely to die before they get a chance to pass their genes on to the next generation.

The theory of natural selection predicts that the most rapidly evolving species will be those whose individuals (1) quickly reach reproductive matu-rity, (2) are under environmental stress, and (3) are prolific in their produc-tion of offspring. Bristlecone pines, with their life span of several thousand years, can afford to grow and reproduce slowly, because there are so few environmental stresses that can affect them; frogs, on the other hand, need to quickly lay millions of eggs to ensure that a few will escape the multitude of hungry predators in the environment. The plant and animal kingdoms are full of thousands of wonderful examples of life forms that have success-fully developed unique traits that enhance their probability of survival (insects that look like twigs, for instance, giraffes that can reach the leaves at the top of a tree, skunks that repel predators with their scent, and northern reptiles with their ability to hibernate through a long winter). Meanwhile, other organisms have developed unique ways of ensuring that their seed sur-vives even if the individual organism perishes: A coconut will float on a hur-ricane's tidal surge, for instance, and the seeds of many fruits will pass unharmed through the intestinal tract of an animal that has eaten them.

Of course, natural selection does not guarantee that an organism will develop such adaptive traits. Clearly, it would be advantageous to modern humans if we could "see" ionizing radiation and thereby avoid it. At present, we might argue that there is not sufficient environmental stress to drive human natural selection in this direction (and indeed I am hard put to postulate the biological mechanism that might result). Beyond this, however, given the time it takes humans to reach puberty and our slow rate of reproduction (roughly twenty-five years per generation), it would take tens of thousands of years for Mother Nature to perform this particular experiment on the human species. If there were a nuclear holocaust tomorrow, or even if the ozone layer disappeared in the next few decades, it is highly unlikely that *homo sapiens* could evolve adaptive mechanisms quickly enough to ensure the survival of the species. In that scenario, the cockroach might emerge as the dominant animal life form on earth, for cockroaches are already somewhat resistant to radiation (an evolutionary head start), and they also reproduce fairly rapidly.

Artificial Selection

Some of the most compelling evidence in support of the theory of natural selection lies in humankind's successes in directing evolutionary change for human benefit. Corn, for instance, evolved from maize plants that were native to the Americas when the first Europeans arrived in the 1500s. Modern corn is quite different from the original maize; its kernels are larger and sweeter, and the plant can no longer grow in the wild, because its seeds do not disperse without the intervention of humans. This evolution from wild maize to corn was extremely rapid, requiring only around one hundred reproductive cycles. In the anthropocentric view, the evolutionary process was one of artificial rather than natural selection, for the newly arrived European farmers consciously chose only the fattest plants from each generation of maize to be used for next year's seeds. From a broader perspective, however, we might view the arrival of the Europeans as a dramatic change in the environment of the native maize, and the evolution from maize to corn as a consequence of this environmental pressure. If you weren't a *fat* maize plant in this new environment, you wouldn't get a chance to reproduce; you'd just be eaten, and that would be that. On the other hand, if you grew very robust kernels on your stalks, the Europeans would save your seed and plant it the following year, and your genes would be passed on to succeeding generations.

In a similar manner, humans have successfully directed the selection process for a broad variety of domesticated animals and food crops. Modern cows, pigs, and most breeds of dogs can no longer survive in the wild. Thoroughbred horses have significantly different traits from wild horses, and farm-grown rice is more nutritious than wild rice. Apricot trees do not grow in the wild, and the broad variety of apples one buys in the supermarket likewise reflects the intervention of humans directing the process of selection.

Carl Sagan describes a fascinating example in which the selection resulted from human superstition and cultural legend rather than from a conscious program to guarantee food for the future.[13] In 1185, on the Japanese Inland Sea, the Heike samurai clan lost a decisive naval battle to the rival Genji samurai. Rather than face capture, great numbers of the Heike warriors threw themselves into the sea. This much is documented history. Upon this historical base a legend has developed: Local fishermen say that the defeated Heike samurai still wander the sea bottom today, in the form of crabs. And in fact, if you travel to the Japanese Inland Sea and pull up a few crabs, you will very likely find some whose carapaces bear the three-dimensional image of the face of a samurai warrior, complete with moustache and chin strap.

The shells of these crabs have grown quite naturally; they haven't been carved or painted by human hands. The human image on its back is therefore part of the genetic makeup of each crab, reflected in the sequencing of the base pairs in its DNA. But what possible link could exist between this internal biochemistry of a crab and the samurai naval engagement of the year 1185?

Clearly, the only possible link is the fishermen. Over the centuries, many millions of crabs have been pulled from the Japanese Inland Sea and eaten. But after 1185, any netted crab that even slightly resembled a Samurai warrior was tossed back into the sea. Over many centuries, this selection by the fishermen resulted in an accumulated survival advantage for those genetic lines of crabs that had Samurai features on their carapaces.

Viewing this process from the perspective of the Heike crabs, the fishermen were part of the crabs' environment, and the selection process was hardly artificial. This was a case of the natural predators (humans) changing their behavior in a way that provided a loophole for the survival of the prey, and the prey (over many generations) increasing its subpopulation of members that had the traits that allowed them to slip through this loophole. The emergence of the Heike crab was rather rapid, as natural evolutionary processes go, but it was not fundamentally different from the process by which electric eels developed their capacity to shock their predators or humans developed their intellectual capacity to plan for the future. Human intervention may change the rate or direction of evolution of a species, but not the fact of evolution.

Clearly, those species that reproduce the fastest and in the largest numbers are statistically most likely to evolve new traits when subjected to environmental stress. Of particular relevance to humans is the case of disease-causing microbes.

Graph *a* in Fig. 4.3 shows how individuals in a population of microbes will vary genetically in their ability to tolerate an antimicrobial drug (an antibiotic, for instance). Suppose that this population of microbes has established itself in a human host, and that the sick person begins taking antibiotics. The first dose of the drug will obviously kill the least tolerant microbes in the population, the second dose will kill the next weakest microbes, and so on. Eventually, after several days of medication, only a small number of microbes remains alive (graph *b*), and the victim of the disease begins to feel pretty chipper again. At this point, a common mistake is to quit taking the drug (perhaps passing the remaining capsules on to a sick friend, or squirreling them away for a future illness). The few surviving microbes, however, are those that were most resistant to the medication. Worse, because the

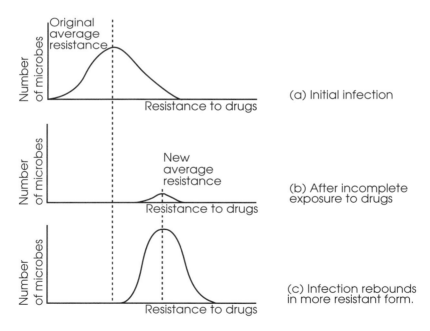

Figure 4.3. Why microbes often develop a resistance to antimicrobial drugs.

treatment has killed off most of their competition, these few strong microbes have been handed an environment conducive to rapid population growth. Now, as shown in graph *c*, the infection rebounds. Note that all of the microbes in this new population are descended from the most drug-resistant microbes in the initial population. Thus the relapse is characterized by an infection that is quite resistant to the initial antibiotic. As a result, the host gets sicker than he or she was initially, and the treatment of the illness requires a longer series of stronger drugs.

Once again, the distinction between natural and artificial selection is . . . , well, artificial. The human organism is a quite natural environment for the multiplication of microbes, and antibiotics in the bloodstream are an environmental stress that drives the evolution of new traits in any resident microbial population. To understand epidemics, we need to keep in mind that microbes are constantly evolving, driven by the same kinds of forces that drive natural selection in all other biological organisms.

Epidemics

Even the healthiest human body harbors hundreds of thriving microscopic life forms: various bacteria, viruses, and often even amoebas (in the mouth). For the most part such microorganisms do not make us ill; in fact, the theory of natural selection suggests that it is a serious survival disadvantage to a microbe to sicken or kill its own host organism. The microbes that are best adapted to the environment of the human body will be fairly benign, because this guarantees their own survival over the long term.

Over the short term, however, wild anomalies can arise. A microorganism can produce a new generation in just hours, compared to the average twenty-five years or so for a human. As a result, within a human lifetime our microbial parasites have incredible numbers of opportunities to develop new traits through mutation and natural selection. Because mutations are random, by far the majority of them result in traits that are in no way beneficial to the microbe's survival. Occasionally, however, just by chance a microorganism acquires a trait that makes it uniquely adapted to thrive with little competition in some part of the human body. When this happens, a new disease is born.

Our current understanding is that it is quite unlikely for a benign human microbe to suddenly mutate into an agent of human disease. What seems to happen instead is this: All animals carry their own uniquely adapted swarms of relatively benign microbes. When one of these microbes mutates, it becomes less adapted to its own host's environment, but it may be more adapted to a different host (a human, for instance). Because we humans often live in close proximity to other animals, we inadvertently give mutated microbes an opportunity to jump hosts. Historical evidence suggests that numerous human diseases originated in this very manner. Smallpox, influenzas, mumps, syphilis, African sleeping sickness, and AIDS, for instance, all seem to have originated with other animals that come into regular contact with humans.

A new microbe that lands in a healthy human will either be quickly gobbled up by the body's natural defenses, or it will overwhelm those defenses in such a way that it can sustain a growth rate greater than 0%. As we have seen, any long-term positive growth rate will eventually lead to repeated doublings of the microbial population. In this case, the microbe's numbers may grow beyond the carrying capacity of the host, and the host will die of the disease.

But even jumping species isn't enough to set off an epidemic in a human population. The other essential ingredient is that the microbe must find a way to transmit itself to a new host before either (1) it kills its original host, or (2) its original host succeeds in exterminating it. If it does not have a way to get from victim to victim in a timely manner, a new disease will quickly fade away on Mother Nature's drawing board.

If you're a microbe, the most efficient way to travel from human to human is to hitch a ride on something humans already do. We humans breathe, for instance, and we also drink water, eat, have sex, stomp around barefooted in gym locker rooms, and expose ourselves to the bites of flies and mosquitoes. These human acts are all fair game for the transmission of infectious microbes from one person to another. Over time, a microbe often develops the ability to affect its host in a way that increases the chance that it will be transmitted to another host: inducing an infected person to sneeze or cough, for instance, to increase the number of airborne microbial spores that float around to be picked up by the next victim, or causing an athlete to scratch his itching toes, leaving fungal spores on the floor of the locker room. In the case of cholera, the microbe increases the rate of expulsion of intestinal waste to such an extreme (as much as 20 liters, or 5 gallons, per day) that it becomes quite likely that it will contaminate the food or water of other prospective hosts. If untreated, cholera is almost always fatal within a day or two; this is not, however, detrimental to the microbe, because it usually succeeds in infecting new hosts very quickly. Outbreaks of cholera are always a danger following any natural disaster that disrupts the delivery of potable water and/or the proper treatment of sewage.

Clearly, a disease will go extinct if it does not spread to new victims. If a disease runs its course slowly, it can afford to spread slowly and still survive, but if it kills quickly, it must also spread quickly. As a result, many of the deadliest diseases are also among the most contagious.[14]

There are, however, two other important factors in this dynamic: population density and mobility. If a population is dispersed and doesn't move about very much, disease-causing microbes will have little opportunity to jump from the sick to the healthy. As a result, every disease that is transmitted from human to human seems to require a certain threshold population density to sustain it. City dwellers contract influenzas (flu) much more often than country folk, whereas in isolated farming communities even the common childhood diseases are rare. Sustaining measles as a disease, for instance, seems to require a population of close to 1 million interacting individuals. In fact, in the early 1800s, most of the French peasants who were drafted into Napoleon's army had never been exposed to measles, and tens of thousands of young conscripts died from this disease before they ever saw an enemy bullet. The same thing happened in the Civil War in the United States, when the Confederacy first assembled its army of young men from small, isolated southern towns and villages: again, thousands quickly died from measles.

In the twentieth century (just four human generations, but millions of generations of the measles microbe), measles has become primarily a childhood disease. This microbe has evolved to a less virulent form that has permitted it to become endemic in our population. Today, if a child is not vaccinated, she will probably be exposed to this disease at a fairly young age, she will get a mild case and recover, and her immune system will forever "remember" the incident and protect her from a recurrence. During this experience, however, she will transmit the disease to enough others to keep the disease alive somewhere else in the general population. Twenty-five years later, her children will pick up the remote descendants of the virus that had sickened her in her own childhood, and the process will continue. The child's own recovery, in other words, has guaranteed the disease a new host at a later date.

This, then, is the ultimate adaptation of a disease-causing microbe. Once it establishes itself as endemic in a population, a disease is unlikely ever to go out of business. The microbe's survival strategy is this: Don't kill too many people; just get them sick enough long enough so that they transmit the disease to sufficient others to keep the microbes around over the long term. Numerous tropical diseases fall into this category: malaria and schistosomiasis, for instance.

As I've already pointed out, however, new diseases are always emerging, and when they first appear they are almost always epidemic rather than endemic. Most of these newcomers are so virulent that they quickly run their course by killing everyone they infect. Legionnaires' disease and Four Corners' disease are two recent examples where a new microbe quickly killed most of its victims and thereby (fortunately) failed to transmit itself into the general population.

AIDS, however, is another story. It kills slowly, it is contagious for a long period before it is symptomatic, and it seems to be invariably fatal. Over the long term, it may be impossible for this disease to survive, because among its victims it targets the very people (the heterosexually active and their fetuses) that it must rely upon to supply new hosts for future generations of the microbe. Yet it's hardly a consoling thought that the AIDS epidemic may run its natural course with the side effect of erasing much of the human species from the planet.

Some will argue, of course, that mutually monogamous couples who also avoid unsterilized injections can survive uninfected amid such a holocaust and become the Adams and Eves of a new and better world. Perhaps, but perhaps not. The AIDS retrovirus mutates at an alarming rate (which is why efforts at developing a vaccine have thus far been unsuccessful). Meanwhile, we have given this microbe an amazingly fertile environment to experiment in, with billions of people worldwide who interact in a variety of nonsexual ways (breathing each other's air, for instance). If the AIDS retrovirus should ever develop an airborne strain, or one that is transmissible through other nonsexual social contact, it will be much too late to take steps to prevent its transfer into that segment of the population that currently views itself as immune by virtue of its monogamous sexual behavior.

Where environments are stagnant, evolution proceeds at a snail's pace, if at all. But where environments change quickly, natural selection always drives some life forms to adapt to the newly emerging ecological niches. As we humans increase in numbers and population density, we also provide our microbial parasites an environment they have never before had available. We are thereby offering the earth's microorganisms an increasing variety of opportunities for developing new human diseases. The laws of probability suggest that under these conditions new diseases will indeed arise in the future, and at an increasing rate.[15] Many, if not most, of these new diseases will begin as epidemics. At least some will be quite lethal.

Is it possible that a future deadly epidemic could become pandemic and wipe the human species from the face of the globe? Yes. Is such a scenario probable? Maybe not, over the short term. However, given enough time and environmental opportunity, relatively improbable natural events become increasingly likely. We do know that over the eons many thousands of life forms have been rendered extinct by causative agents that we can only speculate about. Undoubtedly the culprits in at least some of these instances have been explosive population growth and the transmission of microbial parasites. Yes, we are at considerable risk, and our science is still much too young to give us a general strategy for eliminating that risk.

The Bubonic Plagues

Before the rise of modern sanitation and medicine, cities were the unhealthiest places anyone could possibly choose to live. Until the early 1800s, the death rate exceeded the birth rate in almost every large city in the world. The only way cities managed to sustain their populations during most of human history was by attracting immigration at a rate that compensated for the excessive urban death rates.[16] Some cities, of course, failed to do this and are today populated only by archaeologists.

Naturally, humans would not have flocked to these death traps if there had not been seductive reasons to do so. The attractions were, in virtually all cases, economic: Cities offered employment opportunities and cultural trappings – that is, wealth and prosperity. Cities generated their wealth not by becoming microcosms unto themselves but rather by trading regularly with networks of other cities. Thus evolved an expanding web of trade routes that had the unanticipated consequence of greatly enhancing the opportunities for microbes to infect new populations of densely packed human hosts.

The rise of cities provided human microparasites a new and wonderful ecological niche. Although the evolution of many new human diseases was driven in this manner by the rise of cities, certainly the most terrible to date has been deadly bubonic plague. The most serious outbreaks of this devastating disease occurred in the sixth, fourteenth, and seventeenth centuries, and these three epidemics alone may have killed a total of some 137 million Europeans, in addition to unknown but certainly large numbers of Asians.

The agent of bubonic plague is the bacillus *Yersinia pestis*,[17] which seems to have had at least two natural animal hosts for many thousands of years before human cities arose on the planet. One of these hosts is the common black rat *Rattus rattus*, also known as the house rat, ship rat, or river rat. Another host is the rat flea *Xenopsylla cheopis*, which derives its nourishment by sucking the blood of the black rat. The bacillus multiplies in the bloodstream of an infected rat, eventually infecting the rat's lungs or nervous system and ultimately causing a convulsive death. Until very near its demise, however, the rat tolerates the infection fairly well (maybe even for a year or so), and in this time there is ample opportunity to produce several generations of baby rats. Meanwhile, the rat flea can also tolerate up to a few hundred thousand bacilli in its gut before it begins to regurgitate the bacillus as it tries to feed. Before the bacillus kills a typical rat flea, this tiny insect also has a chance to produce multiple generations of baby fleas.

Before an infected rat dies, the bacillus population in its blood grows so dramatically that no flea on its body can escape infection. When the rat does die, these fleas (which are very sensitive to temperature changes) immediately desert their dead host and look for a new source of nourishment. Although a flea will eat daily if a host is available, it can live for as long as two weeks without eating. In this period, there is a high probability that an infected flea will find a new rat, which, at least eventually, it will infect with the plague bacillus. In this manner, the bacillus is passed back and forth from rat to flea, flea to rat, generation after generation, in an essentially closed community. Some scientists have suggested that this process may have gone on for as long as 1 million years before the first human epidemic of bubonic plague.

We don't know for sure when the bacillus first jumped species and mutated into a strain that could infect humans. A strange epidemic which bore some similarities to bubonic plague swept Athens around 430 B.C., and shortly after A.D. 300 a similar but more severe outbreak killed hundreds of thousands in Rome. Then in A.D. 540 came the great Plague of Justinian, which contemporaries documented well enough to leave little doubt today that this was indeed an outbreak of bubonic plague.[18] This epidemic originated in lower Egypt, moved north to Alexandria on the Mediterranean coast, from there traveled to Palestine, then spread to Constantinople (which at that time was the capital of the Roman Empire), clearly following the major trade routes.

When, in 541, the emperor Justinian returned to his capital from his unsuccessful war in Persia, he found to his horror that his subjects were dying at the rate of up to 10,000 per day. Because even mass graves could not be dug fast enough, Justinian ordered that the roofs be removed from the towers of the city walls to provide places to efficiently stack the corpses; soldiers then poured lye onto these tall piles of bodies to hasten their decomposition. Even this drastic measure, however, was insufficient to dispose of the huge numbers of people who were dying. Eventually all of the city towers were filled with decaying bodies, and the dead had to be loaded into ships, which were then taken to sea and set ablaze. Justinian himself was struck by the disease, and his wife Theodora reigned during his illness. Although he recovered, Justinian never regained his full physical strength and was left with a speech defect for life. By 542, the lives of 40% of the inhabitants of the great city of Constantinople had been claimed by the epidemic, and in fact this may have been the ultimate blow that led to the downfall of the Roman Empire in the Middle East. Meanwhile, the plague did not stop at Constantinople but continued to spread northward and westward, gradually decreasing in intensity until around 590, when apparently the last cases were reported. By that time, the epidemic had destroyed the critical mass of population it needed to sustain itself as a human disease.

This plague was much more lethal to humans than it had been to either rats or fleas, a common characteristic of diseases that jump species. The bacillus had also found new ways to transmit itself directly from person to person, without the need to depend on a flea as an intermediary. The combination of these two factors allowed the bacillus to destroy the very environmental conditions that had contributed to its own population explosion. It was so successful in multiplying, and so lethal to its human hosts, that it put itself out of business.

Although no one would recognize this fact for many centuries, there were three different ways the plague could attack its victims. One was the injection of the bacillus into the blood by the bite of the flea. In this case, a few days would pass before the first symptoms of aches, chills, confusion, and fatigue. Then purple splotches began appearing on the skin, followed by heart flutters, a collapse of the nervous system accompanied by dreadful pain, a brief stage of wild anxiety and terror (sometimes resulting in the so-called "dance of death"), then often death itself. This sequence of symptoms typically took five to seven days. Recovery, however, was not impossible, and, if it happened, the victim was immune from a recurrence for life.

A second form of the disease (which in fact may have been the most common) was pneumonic, where the bacillus was transmitted directly from person to person through the air. Again the incubation period was two to three days, then came a severe bloody cough and death within the next few days. This form of the disease seems to have been fatal in at least 19 of 20 cases. In the third form, an insect bite immediately triggered a violent septicemic reaction by the body, and the victim invariably died the same day (and perhaps within a few hours), before any major visible symptoms appeared. This third course of the disease is still not well understood, and in fact there may have been a carrier other than the rat flea involved here (perhaps a mosquito transferring the bacillus directly from human to human). The contemporary accounts suggest that no one ever survived septocemic plague.

In 1345, eight centuries after the Plague of Justinian, reports began arriving in Europe of a fearful pestilence in the Far East. The next year, as millions were dying in India, this plague expanded along the established trade routes to the West. According to contemporary accounts, the epidemic first entered Europe on a Genoese ship that sailed from the Crimea to the Sicilian port of Messina and docked there in 1347 with a crew of sick and dying sailors. When the disease began spreading into their city, the residents of Messina (much too late to have any effect on their own outbreak of the plague) drove all of the foreign ships back out to sea – eventually to dock elsewhere and accelerate the spread of the epidemic. By the end of the year, the epidemic had spread into northern Italy. By June of 1348, it had engulfed all of Italy, most of France, and the eastern part of Spain. Six months later, it appeared in southern England and Germany. By June of 1349, it had progressed deeper into England, appeared in southern Ireland, engulfed all of France, and was expanding farther into Germany. By the end of 1349, Denmark and Scotland felt its effects. In 1350, the epidemic spread eastward through Scandinavia. Although people living in the countrysides of Europe were often spared, the cities were devastated. Overall, the plague killed roughly one-third of the Europeans alive in 1347.

For city after city, we find gruesome accounts describing the huge number of corpses, the problems of their disposal, and the psychological impact on the survivors. The bubonic plague was by far the most lethal natural disaster of the Middle Ages. This epidemic claimed many times the number of lives of the earlier Plague of Justinian, simply because by 1347 there were many more pockets of high population density where the bacillus could afford the luxury of killing its hosts and still have other hosts to move on to. In 1351, the Vatican estimated that the plague had already claimed 24 million European lives; it is likely that an additional 20 million died before the

European branch of the epidemic ran its course around the end of that century. Meanwhile, the death toll in the Middle East and Asia was probably even higher. Curiously spared, however, was most of Poland. This stroke of good fortune played no small part in that nation's emergence as a major cultural and political power in the late 1300s, a role that it played on the stage of European history for the next two centuries.

The great plague killed much more efficiently than any war humankind has ever known, and it might seem that this high morbidity would have drastically reversed the world's population growth curve. In fact, however, the net population growth rate seems to have been negative for only a few years as the plague swept through. Within two generations after the plague, the world's population count had grown beyond its preplague level; in the six hundred years since then, the world's population has never had another interval of net decline, not even during the two world wars of the twentieth century.

Many contemporary theories arose for the cause of bubonic plague, most drawing upon supernatural or astrological explanations. If anyone indeed proposed that rats and fleas were involved, the proposal apparently wasn't taken seriously. Microorganisms, of course, were unknown in the fourteenth century, so there was no way a complete theory of the epidemic could possibly have been developed at that time.

A little more than two centuries later, in 1665-6, bubonic plague emerged again in western Europe. In London, where this epidemic peaked in the summer of 1665, 2,000 people died each week. This time, however, the plague did not spread as widely, nor was it as lethal. The trend toward less virulent outbreaks of more limited scope has continued ever since: The San Francisco bubonic plague epidemic of 1907, for instance, had 160 documented cases and 77 deaths (and this before antibiotic treatments were developed). Today, in a typical year a few hundred cases will be reported worldwide, with a mortality rate that now hovers around 3%. This suggests that the bacillus has evolved into a form where it is again in relative equilibrium with its global environment, of which human populations have become one integral part.

A Program for Future Epidemics

By definition, disasters are those events that kill, maim, and destroy. To the extent that we provide Mother Nature with increased opportunities to wreak destruction, we shouldn't be surprised to find that she occasionally does so. To put a lot of human eggs in one basket is to assume a greater risk, regardless of how diligently we guard that basket. The risk of disaster is always greatest when a population clusters itself in dense pockets of humanity rather than distributing itself uniformly over the available land area. Clustering, however, is part of our nature as humans, and that much will never change. What we need to recognize is that the more we cluster, the greater the risks we assume.

Even so, meteorological and geophysical events tend to generate fairly localized disasters, affecting only those populations that are unfortunate enough to have concentrated themselves in the affected region. Death clouds from volcanic lakes, monsoon-driven floods, and earthquakes can produce enormous human suffering, but news of such events is cause for sympathy rather than alarm among people living outside the disaster zone. One does not expect a hurricane to generate a worldwide epidemic of tropical storms, nor an earthquake to give rise to a geometrically increasing sequence of earthquakes that circles the globe. Although cities in general are vulnerable to such events, the probability of any single city being victimized, given its fixed location, is usually quite low.

Humans increase their risk when they expand their numbers in those regions that are subject to the more violent ravings of Mother Nature. Objectively, it does not make sense to increase the number of people living in the floodplains of Bangladesh or to allow more immigration into the parts of coastal California exposed to earthquakes. Still, population growth itself is a quite natural phenomenon, and one whose control has hardly begun to yield to our still imperfect social scientific understanding. People live where they live, they move where they move, and they reproduce when they reproduce.

Although it is a serious cause for concern when we see populations expanding into high-risk areas, it is considerably more serious when the risks do the moving. Infectious microbes are by no means bound to specific geographical regions; they move to where the cities are. In developing more rapid and extensive systems of transportation, we humans have provided our present and future microbial parasites a network of extremely efficient channels for getting from one pocket of high human population density to another. Today, a newly evolved infectious disease has the opportunity to spread over the globe at a rate that would leave the fourteenth-century bubonic plague in the dust.[19]

Even ignoring the obvious humanitarian considerations, today's nations can no longer afford to dismiss any epidemic, in however remote a region of the world, as a merely local disaster. To a microbe, one human host is pretty much as good as another, and getting from one of us to another is fairly easy. Our global population has reached a density where the human species is very close to becoming an agar broth in one global petri dish, where the descendants of one opportunistic microbe might ultimately infect the entire human race. If we prefer that this not happen, we need to pay attention. Scientifically.

It's quite true, of course, that even the great epidemics and world wars of history have had little effect on the world's geometrically increasing human census. But although some optimists find great security in this observation, there are also good reasons to find it alarming. If we humans continue to increase our numbers relentlessly, in spite of everything Mother Nature has done historically to reduce our growth rate, and in spite of the obvious limits imposed by the fixed size of our planet, we need to ask what it will take to get the human population ever to level off. The reproductive behavior of our species, unfortunately, seems to be sending us on a course toward a future disaster of unimaginable proportions. And there will always be a sufficient variety of newly evolving microorganisms around to aid in the effort.

To end this discussion on a more upbeat note, the human race does have some things working in its favor. We are, for instance, the only species that has ever achieved any understanding of its own biochemistry, and we've developed a global infrastructure of education and communication that permits new knowledge to be disseminated rapidly. New diseases, wherever and whenever they are observed, do immediately attract teams of scientists to develop the understanding prerequisite to effective containment and treatment. Modern medical science indeed holds great promise for the future of the human race. The unmet challenge lies in ensuring that the population of the human race itself doesn't outrace medical science.

Notes

1 H. Sigurdsson, J. D. Devine, F. M. Tchoua et al., Origin of the lethal gas burst from Lake Monoun, Cameroun, *Journal of Volcanology and Geothermal Research*, 31 (1987), 1-16; H. Sigurdsson, A dead chief's revenge? Scientists now understand the mechanics of the deadly Cameroun gas burst one year ago, but the trigger is still a mystery, *Natural History*, 96 (1987), 44-9.

2 The density of this cloud of carbon dioxide gas would have been even greater due to its cooling upon expansion. This cooling effect also explains the white

cloud, for cooling would have caused water vapor to condense out of the air above the lake. Carbon dioxide itself is transparent.

3   G. W. Kling, M. A. Clark, H. R. Compton et al., The 1986 Lake Nyos gas disaster in Cameroon, West Africa, *Science,* 236 (1987), 169-75.

4   Some sources claim that Bangladesh has two hundred fifty rivers. Although I haven't been able to confirm this particular figure, a study of the maps verifies that the number is indeed very large.

5   I drew the material in this section from a wide variety of secondary sources, including almanacs, atlases, and contemporary news sources. Although some of the specific data may be a little bit fuzzy, the overall picture that emerges is accurate.

6   Variations on these stories have appeared in numerous publications, some apparently going back a century or more. The general principle can be found as early as 1240 in the writings of the Italian mathematician Leonardo Fibonacci. Thomas Malthus is usually credited with first applying these mathematical principles to human population growth, in his 1798 *Essay on the Principle of Population.* Malthus predicted that human population would crash through famine and disease by the early 1800s; when this failed to happen, his writings were generally ridiculed and discredited. A 150-year error, however, is quite insignificant on the time scale of human evolution. It may yet turn out that Malthus was basically right, and that only his projected timing was slightly in error. Malthus's original arguments were revitalized a few decades ago in what was then a bombshell book, P. Ehrlich & A. Ehrlich, *Population, resources, and environment* (New York: Freeman, 1970). Also see note 8 to this chapter.

7   The astute reader may notice that the formula does not give exactly the same doubling time as the arithmetical series calculation. In fact, the formula is more accurate than the series, because the series assumes that all the births take place on the last day of the year, when in fact they take place continuously throughout the year.

8   Some scientific analyses suggest that the earth's maximum long-term carrying capacity is only 1.5 to 2 billion people, which would mean we've already *exceeded* it by a factor of 3; this result was presented at the annual meeting of the Ecological Society of America on August 12, 1994, by Paul Ehrlich of Stanford University and Gretchen Daily of the University of California at Berkeley. A great deal of other thoughtful material has been written on the world's population crisis. Three popularized articles of potential interest to the reader are J. E. Cohen, How many people can the earth hold? *Discover,* Nov. 1992, 114-25; C. C. Mann, How many is too many? *Atlantic Monthly,* Feb. 1993, 47-67; and E. Linden, Megacities, *Time,* Jan. 11, 1993, 28-38.

9   The estimated peak population of 20,000 on Easter Island corresponds to an average density of 122 people per square kilometer. This is about the same as the modern-day population density of Poland or China, and only 15% the density of modern Bangladesh. (See Table 4.1.)

10   D. W. Steadman, Prehistoric extinctions of Pacific island birds: Biodiversity meets zooarchaeology, *Science,* 267 (1995), 1123-31; Paul Bahn, & John Flenley, *Easter Island, earth island* (London: Thames & Hudson, 1992); J. Diamond, Easter's end, *Discover,* Aug. 1995, 62-9.

11   James Watson, Francis Crick, and Maurice Wilkins were awarded the 1962 Nobel Prize in medicine for their work in determining the structure of the DNA molecule.

12   Although natural selection does not specifically program *how* the individual dies, it is interesting to note that the vast majority of deaths of most animals results from oxygen deprivation to the vital organs and tissues (regardless of whether the apparent cause is a heart attack, drowning, cancer, or bubonic plague, for instance). For more details and perspectives on the process of human death, I recommend S. B. Nuland, *How we die* (New York: Vintage/Random House 1994).

13   Carl Sagan, *Cosmos* (New York: Random House, 1980), chap. 2.

14   Examples include anthrax, ebola, and yellow fever. The only reason these diseases remain obscure is that they kill so quickly that they don't get much chance to spread.

15   This point is argued persuasively in L. Garrett, *The coming plague: Newly emerging diseases in a world out of balance* (New York: Penguin, 1994).

16   F. Cartwright, *Disease and history* (New York: Crowell, 1972).

17   There remains a minority scientific opinion that some other but related bacillus may have been responsible for the major historical outbreaks of bubonic plague. Given that the direct evidence is long gone, it is unlikely that this issue will ever be resolved definitively.

18   Accounts and historical analyses of the spread of the bubonic plagues can be found in P. Zeigler, *The black death* (New York: Harper, 1971); William McNeill, *Plagues and peoples* (Garden City, N.Y.: Doubleday, 1976); Geoffrey Marks & William Beatty, *Epidemics* (New York: Scribners, 1976).

19   B. LeGuenno, Emerging viruses, *Scientific American,* Oct. 1995, 56-64.

5

Restless Seas

Action at a Distance

A common feature of meteorological and geophysical disasters is that the site of the destruction may be quite distant from the source that released the energy. I've discussed a number of examples in previous chapters: Lisbon devastated by a tsunami that was generated several hundred kilometers offshore, Mexico City wrecked by an earthquake whose epicenter lay 350 kilometers to the west, and Johnstown scoured by a flood wave that originated 22 kilometers upstream. It is an observed pattern that Mother Nature abhors highly localized concentrations of energy. When a large burst of energy is released, say by an earthquake or a volcano, nature always finds a way to spread it around quickly. Humans quite far removed from a geophysical or meteorological event may therefore still be in grave peril – hardly a reassuring thought. On the other hand, with sufficient understanding of the natural mechanisms of energy transmission, we humans do sometimes have a chance to be forewarned, and therefore forearmed.

In this chapter, we examine one of Mother Nature's tricks for dissipating concentrations of energy: water waves. In the next chapter, we will build upon the present material and see how earthquakes can be understood as a wave phenomenon that has much in common with water waves.

Southern Peru (now northern Chile), 1868

A crew of 235 Americans on the man-of-war *U.S.S. Wateree* had a firsthand experience with a tsunami that many of us would be inclined to dismiss as a fabrication if the documentation weren't so compelling.[1] This ship surfed on the crest of a tsunami and came to rest in the Atacama Desert some 4 kilometers (3 mi) up the coast and 3 kilometers (nearly 2 mi) inland from its initial anchorage. The photograph in Figure 5.1 shows this beached warship;

nearby lay the wreckage of several Peruvian vessels, one of which was found with its anchor chain wrapped around it as far as it would go, indicating that the ship had been rolled over and over by the wave. The *Wateree*, meanwhile, came to rest upright and intact, with the loss of only one crewman, who had been in a small lifeboat at the time. Inhabitants of the coastal cities of Arica and Iquique didn't fare nearly as well; some 25,000 lost their lives to this tsunami and the earthquake that preceded it.[2]

The *Wateree* was one of a class of ships built in the United States at the close of the Civil War to navigate the shallow rivers of the South; for this reason it was flat-bottomed and double-ended, like a canoe. The Civil War ended before the boat could be used for its intended purpose, and it was sent on a cruise to the southern Pacific and the western coast of South America. In August of 1868 the ship was anchored in the harbor at Arica, in what is now northern Chile, while its boilers and engines were being overhauled in preparation for a return cruise to San Francisco. Arica, with a population of

Figure 5.1. The *U.S.S. Wateree*, beached 3 kilometers inland after being struck by a tsunami at Arica on August 13, 1868. (U.S. Navy archives; retouching is on original photograph.)

some 10,000 at that time, was the terminus of the only railroad connecting the coast with Bolivia and therefore had become a center for the machine shops necessary to serve both rolling stock and ships.

The earthquake struck at around 4 PM on August 13, and the initial tremors were felt aboard the ship. Most of the crew ran on deck and watched in horror as the town swayed like the "waves of a troubled sea," then collapsed in a great cloud of dust. The waters of the harbor began to surge and slosh, dragging the international collection of anchored boats in unpredictable directions and smashing some into the cliffs that bordered the harbor. Survivors from the city crowded onto the pier and were quickly swept away by a huge swell in the harbor. The Peruvian man-of-war *America* hastily got up a head of steam and attempted to get out to sea, but to no avail. In the words of Rear Admiral L. G. Billings, recounting the event many years later,

> this time the sea receded until the shipping was left stranded, while as far to seaward as our vision could reach, we saw the rocky bottom of the sea, never before exposed to human gaze, with struggling fish and monsters of the deep left high and dry. The round-bottomed ships keeled over on their beam ends, while the "Wateree" rested easily on her floor-like bottom; and when the returning sea, not like a wave, but rather like an enormous tide, came sweeping back, rolling our unfortunate companion ships over and over, leaving some bottom up and others masses of wreckage, the "Wateree" rose easily over the tossing waters, unharmed.

Billings's account also describes how this returning wave swallowed up a fort and completely washed away its Peruvian garrison and a number of 15-inch cannons weighing several tons apiece. All of this fits the description of a tsunami generated by an undersea earthquake centered fairly close to shore. The *Wateree*'s captain, however, must have been a very cautious man, for he prepared the crew for more to come. Returning to Billings's words,

> It had now been dark for some time and we knew not where we were, the absence of the usual beacon and shore lights adding to our confusion. About 8:30 p.m. the lookout hailed the deck and reported a breaker approaching. Looking seaward, we saw, first, a thin line of phosphorescent light, which loomed higher and higher until it seemed to touch the sky; its crest, crowned with the death light of phosphorescent glow, showing the sullen masses of water below. Heralded by the thundering

roar of a thousand breakers combined, the dreaded tidal wave was upon us at last. Of all the horrors of this dreadful time, this seemed the worst. Chained to the spot, helpless to escape, with all the preparations made which human skill could suggest, we could but watch the monster wave approach without the sustaining help of action. We could only grip the life-line and await the coming catastrophe.

The ship came to rest in the sand 3 kilometers inland and just 60 meters short of being dashed against a cliff. The next morning, the ship's navigator measured the high water mark on the adjacent mountain at 14.3 meters (47 ft) above the sand, "not including the comb" of the wave (although it is not clear how the latter was established). The U.S. Coast and Geodetic Survey has since estimated the tsunami's height at approximately 21 meters (70 ft) at the time it struck the *Wateree*. In Arica, this wave swept away heavy pieces of machine shop equipment and even complete railroad trains, including their locomotives, leaving behind no trace.

At Iquique, 193 kilometers (120 mi) to the south, the receding wave uncovered the bay to a depth of 7.3 meters (24 ft) and returned with a 12-meter (40-ft) wave crest that engulfed the city. The earthquake that generated the wave permanently raised sections of the shoreline between Arica and Iquique as much as 6 meters (20 ft). The tsunami was recorded at the Sandwich Islands, 5,580 nautical miles distant, only 12 hours and 37 minutes after it struck Arica. In order for the wave to travel this far this fast, the average wave speed had to have been some 800 kilometers per hour (500 mi/h)! Modern jetliners don't travel much faster.

Every attempt to base a scientific interpretation on eyewitness accounts is, admittedly, problematic. If we take Billings's account at face value, however, it would appear that he and his fellow crewmen experienced not one series of tsunamis, but two. There was too much of a time lag, more than 4 hours, between the earthquake and the last great wave for the two events to have had a simple connection. A more reasonable scenario is this: One of the earthquake's aftershocks (Billings's account mentions a number of these) may have triggered an underwater landslide off the continental shelf, perhaps offshore of the mouth of the Uuta River, where silt could easily have been accumulating for centuries. It would have required only a relatively mild aftershock to produce a devastating tsunami in this manner, particularly if the accumulation of silt had already been rendered unstable by the prior earthquake. The possibility of such a delayed effect suggests that it may not be prudent for a population to consider the seas safe from tsunamis until many hours, or even days, after an earthquake.

And what happened to the *Wateree* and her crew? The stranded sailors were rescued three weeks later by the U.S. frigate *Powhatan*, which was making a scheduled stop. The *Wateree*, though undamaged, was hopelessly mired in the sand much too far from the sea for there to be any hope of refloating the ship. It was sold at auction to a hotel company, then used successively as a hospital and warehouse, finally succumbing to destruction by artillery during the Peruvian–Chilean War. The remains of its iron ribs have since disappeared into the shifting desert sands of what is today northern Chile.

Describing Waves

A wave is a disturbance that transfers energy through a medium (i.e., a "carrier") in such a way that the medium itself remains intact after the wave has passed.[3] The medium for water waves is water; for sound waves it is the air; and for earthquake waves it is the rock and soil that make up the earth. Although electromagnetic waves such as visible light can travel through a vacuum, all other types of waves need a physical medium to carry them.

Waves arise when the medium is deformed by a transient infusion of energy. Toss a stone into a pond, and you've injected energy into a small portion of the pond's surface. The result is not merely a splash; if you watch closely you will subsequently see a series of concentric circles of wave crests and troughs radiating outward from the initial point of impact. These waves transfer the initial energy to the banks of the pond, causing a small amount of erosion. When everything calms down again, the pond (the medium) is just as it was before, but the banks may be slightly different.

The diagram in Figure 5.2 shows a water wave whose source is the sudden upheaval of a portion of the seabed. The impulsive motion of the ocean floor raises a low, broad hump of water on the surface, extending over an area roughly comparable in size to the undersea earthquake source. A hump of water, however, is highly unstable, and it responds to the force of gravity to bring it back to its equilibrium level. As it falls, this mass of water gains enough momentum to overshoot its original equilibrium position and create a trough. This in turn forces the adjacent water upward, thus transferring energy outward along the water surface. The adjacent water behaves similarly, and soon the entire surface is covered by a pattern of undulating crests and troughs that carry energy away from the site of the initial disturbance.

In sketching Figure 5.2, I was forced to leave out the most important characteristic of the wave: the fact that it is *moving*. What I have shown is

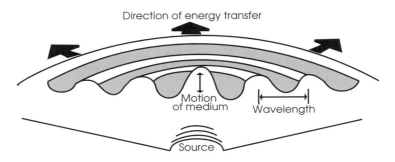

Figure 5.2. A water wave generated by an idealized local disturbance in the sea floor.

only a snapshot of one instant in time shortly after the seabed disturbance took place. An instant later, the pattern of crests and troughs will be shifted horizontally. The visual image, if observed in real time, is that of a surface pattern continuously moving outward from the source. Each drop of water, however, is actually moving in an oval path, as can easily be demonstrated by dropping a floating object on the wave and watching its motion. The wave, in other words, does not involve any net transfer of the water itself; its only long-term effect is a net transfer of *energy*.

Scientists use a standardized nomenclature to describe the properties of waves. The horizontal distance between two consecutive crests of a wave (or two consecutive troughs, for that matter) is called the *wavelength*. Wavelength is measured in conventional distance units, such as meters, feet, and miles. A wave's wavelength is typically on the same order of magnitude as the size of the source (by "order of magnitude" I mean within a factor of 10 or so). Large sources such as undersea earthquakes may produce waves with wavelengths of several hundred kilometers, while smaller sources, such as localized wind gusts over deep water, may generate wavelengths of only 60 to 100 meters.

In addition to its length, a wave will also have a *period*, defined as the time it takes one wavelength of the wave to pass a fixed point. The period is measured in conventional time units, such as seconds, minutes, and hours. To measure the period of a water wave, one can simply toss an empty beverage can onto the wave and clock the can's bobbing motion. If, for instance, it bobs up and down through one complete cycle in 8 seconds, then 8 seconds is also the period of the wave. This is also, of course, the period of oscillation of the can itself, as defined in the section on dynamic loading in Chapter 3.

A third quantity of interest is the *wave speed,* the speed at which the wave energy travels forward. For instance, if a wave crest travels 3,000 meters in 300 seconds, the wavespeed is 10 meters per second. Similarly, a wave that travels 6,300 miles in 12.6 hours has had an average wave speed of 6,300 ÷ 12.6, or 500 miles per hour. Such results must, of course, always reflect the appropriate measurement units.

Repeated studies have shown, however, that wave speeds are predictable enough that they seldom need to be measured directly. In other words, if a certain type of wave travels through a specified medium, the wave speed has no choice but to assume a certain value. Table 5.1 lists some representative values for different waves and media. In practice, one can calculate adjustments to these data to account for changes in temperature, pressure, and other factors that sometimes affect the medium's physical properties. In general, factors that increase the medium's ability to restore itself to equilibrium will increase the wave speed, whereas increasing the medium's density (its inertia per unit volume) will decrease the wave speed.

A wave's wavelength, period, and wave speed are closely interrelated. In fact, the following formula applies to all types of waves (not just water waves):

$$\text{Wavelength} = \text{period} \times \text{wave speed.}$$

For instance, if a tsunami has a wave speed of 500 miles per hour and a period of 0.50 hours, the wavelength is 0.50×500, or 250 miles. Similarly, a sound wave with a period of 0.02 seconds traveling at 343 meters per second has a wavelength of 0.02×343, or 6.9 meters. (Note again that the measurement units must be consistent for this formula to give a meaningful answer.)

In addition to the quantities I've just described, a wave also has another obvious feature: its height. Even the most casual observer will notice that wave damage to shoreline structures is greater when the waves are taller. Is this a linear effect? That is, if we double a wave's height, does it deliver double the energy? The answer is no; the effect is much more dramatic than this. All other factors being equal, doubling a wave's height actually increases its energy by a factor of 4, and tripling the height increases the energy by a factor of 3^2, or 9 times. Moreover, the wavelength is also a factor, for long waves are more destructive than short ones. A precise formula can be written, but for our purposes here the following will be adequate:

$$\text{Wave energy } \textit{is proportional to } (\text{wavelength}) \times (\text{height}).^2$$

Table 5.1. *Wave speeds of some types of waves in different media*

Type of wave	Medium	Other conditions	Wave speed (m/s)
Light	Vacuum		299,792,458
Light	Water	20 °C, yellow sodium light	224,842,000
Light	Diamond	Yellow sodium light	124,010,000
Sound	Air	0 °C	331.45
Sound	Air	20 °C	343.3
Sound	Water	Pure, 20 °C	1,484.7
Sound	Water	Sea, 20 °C	1,519
Sound	Granite	20 °C	6,000
Water wave	Deep water	10-m wavelength, depth > 200 m	4.0
Water wave	Deep water	40-m wavelength, depth > 800 m	7.9
Water wave	Shallow water	Depth 5 m, wavelength > 10 m	7.0
Water wave	Shallow water	Depth 20 m, wavelength > 40 m	14.0
Water wave	Shallow water	Depth 100 m, wavelength > 200 m	31.3
Tsunami	Open ocean	Depth 5 km, wavelength > 10 km	221
P-wave[a]	Bedrock	Depth 0 km	5,400
P-wave	Bedrock	Depth 20 km	6,250
S-wave[a]	Bedrock	Depth 0 km	3,200
S-wave	Bedrock	Depth 20 km	3,500

Note: To convert units: 1 m/s = 3.6 km/h = 3.28084 ft/s = 2.236936 mi/h.
[a]P-waves and S-waves are the two principal types of earthquake waves, discussed in Chapter 6.

Proportionality statements like this permit us to compare two situations without the mathematical complications of more detailed calculations. For instance, suppose that one wave (wave A) has a wavelength of 60 meters and a height of 3.5 meters, whereas a second wave (wave B) has a wavelength of 5,000 meters and a height of 7.0 meters. Clearly, wave B packs more energy. The question is how much more. We can find out by computing the product of the wavelength and the square of the height for wave B, then dividing by the corresponding quantity for wave A. The result is 333 (a pure number with no units). What does this mean? It means that wave B carries 333 *times as much* energy as wave A. Wave B is therefore capable of causing roughly 333 times as much damage to shoreline structures as wave A. Notice that we reach this conclusion without knowing the precise amount of energy in either wave. Proportional reasoning of this type is a very powerful analytical tool, in that it tells us so much for so little.

I need to clarify one other point about this particular proportionality relationship: It refers only to comparisons between waves that are similar in all respects *except* wavelength and wave height. If the two waves have different widths (e.g., one strikes 100 meters of coastline while the other strikes 1,000 meters of coastline), or if one wave is breaking while the other is not, the relationship no longer holds. Nor can we use this relationship to compare two different kinds of waves, say a water wave and a wave through the ground. Even with a proportionality relationship, we can compare only those events that are fundamentally similar.

Energy Transformations

Because the idea of "energy" is so ubiquitous in our modern culture, I felt safe in using this term in the preceding sections without formally defining it. Before moving on, however, we should briefly examine this concept a bit further.

Scientists use the term *energy* to describe a system's capacity to generate moving forces. The international measurement unit for energy is the *joule* (J), defined as the capacity to generate a 1-newton force that acts through a distance of 1 meter. In most of the phenomena responsible for natural disasters, the energy transfer is many megajoules (millions of joules) or even gigajoules (billions of joules). The release of 1 megajoule of energy can generate a force of 1 million newtons acting through a distance of 1 meter, or 500,000 newtons acting through 2 meters, or 100,000 newtons acting through 10 meters, or any other combination of force and distance that gives the same multiplicative product.

Knowing a system's energy does not tell us exactly what the system will do. Rather, it tells us what the system is *capable* of doing. It was only around 150 years ago that scientists realized that every physical event can be described as an energy transformation, and that an energy paradigm can be used to distinguish between possible and impossible physical events. The fundamental principle, which is usually referred to as the "law of conservation of energy," can be stated as follows:

Energy is never created nor destroyed; it is only changed from one form to another.

What energy forms are we talking about here? Many have been identified, but they all fall into two general categories: *kinetic energy* (energy of motion), and *potential energy* (stored energy). Kinetic energy is associated with things that move: wind, streams of water, waves, windborne debris, electrical currents, and even individual moving molecules. Rather than listing the many formulas that have been developed for calculating the energy of motion for all of these cases, let me simply point out that such formulas do exist, and that they tend to be fairly accurate if used properly. A system does not need to be moving, however, to be capable of releasing energy. Many stationary systems also store energy that is capable of being liberated: water behind a dam, compressed gases in a volcanic dome, electrically charged thunderclouds, or a tank of gasoline, for instance. Once again, tested formulas exist for calculating most of these different types of potential energy. The only prerequisite is that the relevant measurement data be available on which to base such calculations.

With these ideas in mind, let's return to Figure 5.2. The law of conservation of energy tells us that the waves on the water's surface can have no more total energy than the energy released by the source – here, an underwater earthquake. In fact, only a portion of the source energy is coupled into the water waves themselves; some of the released energy travels through the seabed in the form of earthquake waves, a little is converted to heat, and some may enter the atmosphere as an acoustical (sound) wave. It is the *total* energy of the water waves, plus the earthquake waves, plus the heat, plus the acoustical waves that adds up to the energy released at the source. If it someday becomes possible to predict the energy released by an earthquake before it occurs, it will also be possible to establish upper limits on the potential damage caused by the resulting earthquake waves and/or tsunamis. Clearly, such a scientific breakthrough would be of great value to disaster planners.

Of course there is another link that must be made here, and that is the effect of distance from the source. It's quite obvious that the farther we are from a geophysical event, the less damage we experience. This follows from simple geometry: As a circular wave front expands in all directions away from its source, its energy is spread over an increasingly larger region, and the wave height diminishes. The total wave energy is still there, but it is no longer as concentrated in space. Because our human structures occupy fixed amounts of space, they will absorb only a small portion of the energy of a wave that has had the opportunity to expand before it hits.

Sometimes, however, there are unpleasant surprises. It is possible for waves to be deflected and focused by obstacles in their path in such a way that the energy becomes more concentrated, rather than less. In April of 1930, for instance, waves 4 meters (13 ft) high with periods of 20 to 30 seconds dislodged stones weighing up to 20 tons from a breakwater at Long Beach, California, while there was only minimal wave activity along adjacent shores to the north and south and the waves at sea were only a half meter in height. It was seventeen years before an explanation was found: An underwater hump 6 kilometers away acted as a lens to focus waves coming from a particular direction, and when such waves happened to have a 20- to 30-second period, the focal point of this lens coincided precisely with the location of the breakwater.[4] The actual source of the destructive waves was apparently thousands of kilometers distant. Given the difficulty of analyzing this event even after it had happened, it is unlikely that any amount of prior study and computation by the engineers would have predicted its occurrence. Even with phenomena whose mechanisms we understand fairly well, Mother Nature still seems to retain the capacity to catch us by surprise.

The Superposition of Waves

So far, we have spoken only of the ideal wave, which has a well-defined period, wavelength, wave speed, and height. Such waves can easily be created in wave tanks in the laboratory, and their properties are well understood. Nature, however, does not control its variables as diligently as do scientists; if you look at the surface of a body of water in a windstorm, or a seismographic recording of an earthquake, it is immediately apparent that the wave action has no single, well-defined wavelength or period. Rather, the motion of the medium seems to be an almost random hodgepodge.

In 1822, the French mathematician Jean Baptiste Fourier discovered that every complex wave form can be viewed as the sum of a series of simple ideal waves, each with its own well-defined period, wavelength, and height. This discovery turns out to be more than a mathematical sleight of hand; in nature, complex wave forms do result from the superposition of multiple simple waves. The diagrams in Figure 5.3 shows how four simple waves combine to create a more complex wave when they are added together.

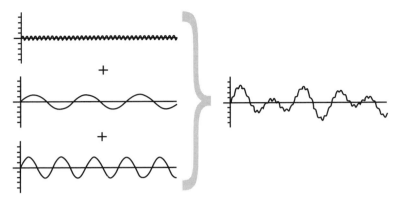

Figure 5.3. The superposition of waves. Complex waveforms result from the superposition of simple waves with well-defined periods, wavelengths, and heights.

When we say that waves are "added," or "superimposed," we mean that physically one is riding on top of the other. However, because waves have troughs as well as crests, it sometimes happens that a cresting wave fills in the trough of an underlying wave, so that all wave action disappears at that point in space. This phenomenon is called *destructive superposition*. But if two waves combine to produce no wave, where does the energy go? The answer lies in the fact that waves have extension in both space and time. Any energy that disappears at one place must show up somewhere else, in particular at the nearest places where the superimposed waves have coinciding crests and troughs. Where this happens, we say we have *constructive superposition*. The effect is shown in the diagrams in Figure 5.4.

The surface of the ocean at any instant results from the superposition of many waves from many sources, some of them possibly quite distant (e.g., last week's storm off the African coast, a closer squall, the local wind, etc.). These waves cross while traveling in different directions, and occasionally for a few moments they may superimpose to produce a calm spot. On the other hand, they may also combine to produce a momentary whopper of a wave that appears to rise from nowhere and then quickly disappears; cases of such "rogue" or "phantom" waves as high as 40 meters (130 ft) have been reported. A rogue wave can be a serious hazards to a large ship such as a tanker, because it can buoy up the center of the ship while leaving the bow and stern hanging out in space. In this configuration, most ships will quickly break in two. Fortunately, the very largest rogue waves are relatively rare, and none seems to occur close enough to shore to threaten shoreline struc-

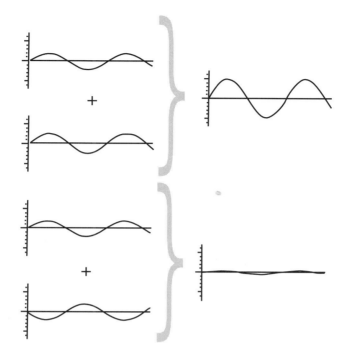

Figure 5.4. Constructive and destructive superposition of waves.

tures. On the other hand, more than a few major oil spills seem to have been produced by this natural phenomenon.

Tides

All water waves, including tsunamis and storm waves, are superimposed upon the natural rise and fall of the tides. A hurricane or tsunami that makes landfall during a high tide will therefore present a greater threat than one that arrives when the tide is low. Although in themselves the tides are gentle and predictable, through the mechanism of superposition they can significantly aggravate or mitigate the shoreline effects of other waves that ride upon them.

As the diagram in Figure 5.5 reveals, tides result from a combination of three factors: (1) The Moon's gravity exerts a stronger pull on the ocean waters that face it than on the waters that face away from it; (2) the Sun's

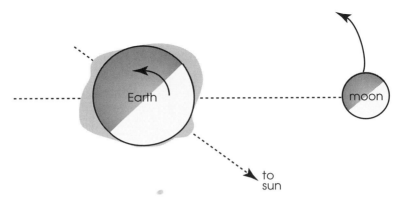

Figure 5.5. Tidal bulges in Earth's oceans result from the gravitational influences of the Sun and the Moon.

gravity also affects the oceans differentially (although to a smaller extent);
and (3) Earth rotates on its axis. The result is two pairs of tidal bulges in
Earth's oceans, one pair associated with the Moon and a much smaller pair
associated with the Sun. These bulges behave like very long waves that
sweep around the globe as Earth rotates under them; in other words, at any
given point on a shoreline the rise and fall of the tide will be perceived as a
very long-period wave. How long? Given that Earth rotates once in 24
hours, it might be expected that the two lunar-induced wave crests will be
separated by an interval of precisely 12 hours. This analysis, however,
neglects the fact that the Moon is also orbiting Earth in the general direction of Earth's rotation. As a result, consecutive high tides average more like
12 hours, 15 minutes apart. Although a simple analysis might also suggest
that a tide should be highest when the Moon is directly overhead, if you go
to a shore, you may observe that a high tide in fact arrives in advance of the
Moon. The explanation again lies in the rotating Earth, which drags the tidal
bulges along with it in its direction of rotation.

Although tidal bulges in the open oceans average only around 30 centimeters in height (0.3 m, or about 1 ft), at coastlines they are often considerably higher. The relatively shallow continental shelves act as wedges as
Earth's rotation drives them under the tidal bulges, and this helps lift the
tides at shorelines. How high? It depends on where you are and when you're
there. On small midocean islands, tidal fluctuations are seldom greater than
in the open ocean, about one-third of a meter. The other extreme can be seen
in the Bay of Fundy where the funnel-like shape of the shoreline and the
wedge of the sea bottom channel the tidal energy, and high and low tides

often differ by as much as 12 meters (40 ft) in the vertical. On the more gently sloping shorelines of this bay, these tides sweep inland as much as 1 kilometer or more, whereas in other places they reverse the flow of rivers. These are dramatic events to watch unfold, and one that seaside vacationers continually need to be warned about. Clearly, the tides at shorelines involve complex interactions between the sea and the topography of the local basin.

Tides vary not only with location, but also with the calendar. Table 5.2 shows a tide table for Apalachee Bay, Florida, for a portion of the month of July 1994. The highest high tides (128 cm) occurred on July 22-3, which was also the date of the full moon. This is no coincidence, for in referring back to Figure 5.5 we expect the highest of high tides to occur when the solar and lunar tidal bulges are superimposed. This happens twice each month: once when the Sun and Moon are on nearly directly opposite sides of Earth (full moon), and again roughly two weeks later when the Sun and Moon are aligned on the same side of Earth (new moon). These superimposed solar and lunar tides are referred to as "spring tides," while the lower tides occurring when the Moon is in a quarter phase are called "neap tides." Even during a spring tide, however, Earth's two tidal bulges may be unequal. This shows up in the tide table of Table 5.2 as the series of consecutive high tides, beginning July 20, of 122, 101, 125, 104, 128, 107, 128, 107, 125, then 110 centimeters. In other words, every second high tide was higher than the intermediate high tide. This inequality of consecutive tides happens when the Moon does not lie in the geometrical plane of Earth's equator (which is most of the time). When the Moon's orbit does align with the plane of Earth's equator, consecutive high tides are approximately equal. This happens twice a year, but on specific days that vary from year to year.

Despite all of these complexities, today it is possible to predict the tides to a high degree of accuracy for any future day or time of interest, at virtually any portion of any coastline. In fact, the data in annual tide books is typically calculated two years in advance. The tides are driven by an astronomical clockwork, and surprises are unlikely, if the mathematical analysis is done correctly. Still, the ubiquitous tiger tails do get into this clockwork, for the shorter waves that ride upon the predictable tides are often notoriously unpredictable.

Waves in Deep and Shallow Water

"Deep" and "shallow" are, of course, relative terms; what is deep to a human may be shallow to a ship. Similarly, what is deep to a short-wavelength wave

Table 5.2. *Tide table for Apalachee Bay, Florida, for the month of July 1994*

	Time	Height			Time	Height	
	h m	ft	cm		h m	ft	cm
10	0322	3.5	107	21	0119	3.3	101
Su	0841	1.3	40	Th	0620	1.7	52
	1431	4.1	125		1232	4.1	125
	2126	−0.2	−6		1939	−0.5	−15
11	0352	3.6	110	22	0201	3.4	104
M	0921	1.1	34	F	0712	1.5	46
	1511	4.1	125		1321	4.2	128
	2156	−0.1	−3		2021	−0.4	−12
12	0422	3.7	113	23	0237	3.5	107
Tu	1004	1.0	30	Sa	0758	1.3	40
	1556	4.0	122		1405	4.2	128
	2229	0.1	3		2057	−0.2	−6
13	0453	3.7	113	24	0310	3.5	107
W	1053	0.9	27	Su	0841	1.2	37
	1645	3.7	113		1446	4.1	125
	2305	0.4	12		2130	0.0	0
14	0528	3.7	113	25	0340	3.6	110
Th	1148	0.8	24	M	0921	1.1	34
	1744	3.4	104		1525	3.9	119
	2345	0.8	24		2159	0.3	9
15	0608	3.7	113	26	0408	3.5	107
F	1255	0.7	21	Tu	1001	1.0	30
	1857	3.0	91		1603	3.6	110
					2226	0.6	18
16	0033	1.2	37	27	0434	3.5	107
Sa	0656	3.7	113	W	1043	1.0	30
	1414	0.6	18		1643	3.3	101
	2031	2.8	85		2252	0.9	27
17	0132	1.6	49	28	0501	3.5	107
Su	0758	3.7	113	Th	1128	1.1	34
	1539	0.4	12		1729	3.0	91
	2210	2.8	85		2320	1.2	37

Table 5.2. *(cont.)*

		July						
	Time	Height				Time	Height	
	h m	ft	cm			h m	ft	cm
18	0244	1.9	58	29		0529	3.4	104
M	0914	3.7	113	F		1224	1.2	37
	1655	0.1	3			1827	2.7	82
	2331	2.9	88			2354	1.5	46
19	0405	2.0	61	30		0604	3.3	101
Tu	1031	3.8	116	Sa		1339	1.2	37
	1759	−0.2	−6			1952	2.5	76
20	0031	3.1	94	31		0039	1.8	55
W	0518	1.9	58	Su		0653	3.2	98
	1136	4.0	122			1513	1.2	37
	1852	−0.4	−12			2141	2.4	73

may be shallow to a long-wavelength wave. The underlying physical consideration here is whether or not the wave "feels" the bottom. If the depth of the water is less than about half a wavelength, the wave's motion extends all the way to the sea floor, and we consider the water to be shallow. If the water is deeper than about 20 wavelengths, the bottom remains undisturbed by the passage of the wave, and we consider the water to be deep. Between these two extremes is a transitional region that does not lend itself to an Aristotelean "either-or" dichotomy. A wave traveling from deep to shallow water transforms continuously from a deep-water wave to a shallow-water wave.

The distinction is important, because deep-water waves and shallow-water waves travel at different speeds. For deep-water waves, the wave speed depends on the wavelength:

$$\text{Wave speed (in m/s)} = 1.249 \times \sqrt{\text{wavelength (in m)}}. \quad \text{[deep water]}$$

This equation predicts, for instance, that in deep water a wave with a 20-meter wavelength will propagate at 5.59 meters per second (12.5 mi/h), whereas a wave with an 80-meter wavelength will travel at 11.2 meters per second (25.0 mi/h). Given our earlier principle that wavelength is always equal to wavespeed times period, the periods of these two waves are, respectively, 3.58 seconds and 7.16 seconds. In this relationship, then, we see the reason why the surface of a deep body of water often looks so confused. If multiple waves are present (and they almost always are), the longer waves will keep overtaking the shorter ones, and the patterns of superposition will continually change. The eye perceives a state of chaos, rather than the underlying rhythm of the individual waves.

For shallow-water waves, which "feel" the bottom, the wave speed no longer has anything to do with the wavelength. Instead, another variable enters: the depth of the water. The following equation describes this interaction:[5]

$$\text{Wave speed (in m/s)} = 3.132 \times \sqrt{\text{water depth (in m)}} \quad \text{[shallow water]}$$

For instance, if a 20-meter wave and an 80-meter wave simultaneously enter a basin whose depth is 10 meters, this second equation predicts that both will travel forward at a speed of 9.90 meters per second (22.1 mi/h). Because in shallow water these two waves have identical wave speeds, their superposition pattern is relatively stable, and an obvious string of wave crests will be observed traveling toward shore. Thus, out of the apparent chaos of the deep water, we see emerge a clearly identifiable series of shore-bound waves.

Every wave impinging on a coastline eventually becomes a shallow-water wave. Once this transition has been made, the wave speed continuously decreases as the water depth decreases. In water 10 meters deep the waves travel at 9.90 meters per second (22.1 mi/h), but when the water shallows to a 2-meter depth (6.5 ft) the wave speeds decrease to 4.43 meters per second (less than 10 mi/h). Although it might seem that this decrease in wave speed ought to render a wave more benign, in reality just the opposite happens: As the water gets more shallow, the wave grows in height.

Why? Because in a sloping basin, the leading portions of a wave encounter shallow water before the trailing portions of the same wave. This creates a condition where the trailing section of the wave is traveling faster than the section of the wave immediately in front of it. When this happens with cars on a busy freeway (in a fog, for instance), we get a chain-reaction collision where the faster cars pile into the slower cars in front of them. When it happens with a wave, the faster portion of the wave piles on top of the slower portion in front of it. None of the wave's energy disappears as it grows in this manner. What does happen is that the wave's energy becomes concentrated in a higher but narrower wave crest as it approaches the shore. This process is shown in the diagram in Figure 5.6. Remember that here again I have been forced to leave out the most important characteristic of the wave: the fact that it is *moving*.

As waves enter shallow water, then, they slow down and grow in height. Two additional things also happen: Their wavelength decreases, and they break. Let's examine these phenomena one at a time.

We saw earlier in this chapter that wavelength = wave speed × period. A wave's period (the time between crests) stays relatively constant throughout the life of the wave, regardless of its adventures. As a result, the wavelength shrinks when the wave speed decreases. For example, if the wavelength is 21 meters when the wave speed is 6 meters per second, a decrease to a 2 meters per second wave speed will shrink the wavelength to 7 meters. Throughout this process, the period stays fairly constant, at 3.5 seconds. The shorter wavelength corresponds to the "piling up" effect just described, which acts to increase the wave height when the wave slows down.

Can a wave ever speed up? Yes. If a wave is generated in shallow water and travels into deeper water, its wave speed increases, its wavelength increases, and its height decreases. All of the processes described so far, in other words, work equally well in reverse.

"Breaking," however, never works in reverse; it is one of nature's one-way streets. A wave breaks when there is no longer enough water forward of the crest to sustain the shape of the wave. This situation occurs naturally on gently sloping beaches, when the water becomes shallower than the wave is high; it also occurs at artificial breakwaters, where a wave encounters an abrupt transition from water to rock or concrete. When a wave breaks, the

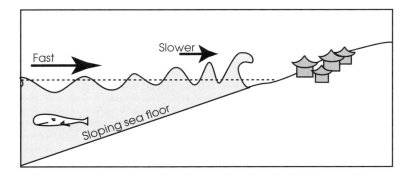

Figure 5.6. Water waves grow in height as they enter shallow water.

water at the wave crest that is following its usual orbital motion is thrown forward into empty space. If the wave is very high, the kinetic energy released by this process is capable of causing great damage to any obstacle in its path. It is not only human constructions that are vulnerable; entire beaches have been known to be washed away overnight by heavy breakers.

As usual, Mother Nature does not keep things simple. It is possible for a wave to scour a beach without breaking, and it is possible for a wave to break without coming anywhere close to shallow water. As a rough rule, we can figure that a wave will break when its height exceeds about one-seventh of its wavelength. Very long waves, then, can sometimes inundate shorelines without breaking; in fact, observers of some tsunamis have reported that there was no breaker, but that the ocean simply "sloshed" over the shoreline to a great height.

Most water waves are, of course, generated by wind. The higher the waves, the more efficient is the energy transfer from wind to water that causes the waves to grow further. Apparently the highest recorded wave experienced by a ship at sea, as reported by the officers of the *U.S.S. Ramapo* in 1933, measured 34.1 meters (112 ft) in height. This huge wave presented no real danger to the ship, because the wave was not breaking and the crew was skilled enough to keep the ship pointed into the wave. Computer simulations suggest that it is theoretically possible for a storm wave to grow to a height of 67 meters (219 ft)!

Normally, a wind-driven wave reaches a height of one-seventh its wavelength long before it attains such monumental proportions, and at this point it dissipates its energy as a frothing, white-capped breaker. It is then that the wave becomes dangerous. Waves present little threat as long as they continue "waving." But when a wave terminates in a breaker, or even in a large swell on a shoreline, a devastating amount of energy is transmitted to whatever lies in its path.

Tsunamis

Nonscientific sources often refer to these devastating waves as "tidal waves," much to the chagrin of oceanographers who keep shouting that the phenomenon has no connection whatsoever with the tides. More appropriate terms are "seismic sea wave" and "tsunami," the latter a Japanese word that translates as "harbor wave" (reflecting the fact that these waves are innocuous until they approach shorelines). A list of notable historical tsunamis is given in Appendix A.

Tsunamis are impulsively generated; they arise when a large amount of energy is rapidly pumped into a body of water over a large area. High winds, no matter how intense, cannot create a tsunami, because wind does not transfer energy to the sea in an abrupt impulse. Tsunamis are generated by sudden releases of energy in undersea earthquakes, explosions of sea-level volcanoes, and by undersea landslides at the continental shelves. In a few bizarre cases, they have also been caused by large coastal landslides that fell into the sea.

Tsunamis have very long periods, typically 20 minutes to 1 hour, and wavelengths measuring hundreds of kilometers. Because the crests are so far apart, observers often report only a single giant wave. There is always, however, a whole series of wave crests in any tsunami, and the first crest is not necessarily the largest. One should never make the mistake of assuming the show is over when the first wave crest recedes.

Tsunamis carry some 1% to 10% of the energy released in the event that causes them; typical tsunami energies are in the range of 10,000 gigajoules (ten thousand *billion* joules) to 100,000 gigajoules.[6] Because a tsunami's wavelength is so long, however, this tremendous energy results in a modest wave height of only 1 meter or so in the open ocean. A ship in deep water will not even notice a tsunami passing under it, because the wave's rise and fall over a period of up to 1 hour may only be about the height of a desktop. On June 15, 1896, a devastating amount of energy raced undetected in this manner beneath several fishing fleets off the coast of Japan; when these boats returned to their home ports, they found their villages completely washed away along 483 kilometers of coastline. The official census tallied 26,975 dead, 5,390 injured, 9,313 houses destroyed, some 300 large ships stranded, and around 10,000 smaller fishing boats crushed or sunk in the harbors.[7] Yet the fishermen in deep water had no hint that all of this energy had passed beneath them.

As we saw in Figure 5.6 and the accompanying discussion, shallow water decreases a wave's speed of propagation, which in turn shrinks its wavelength and increases its height. If an open-sea tsunami has a height of 1 meter when its wavelength is 400 kilometers and its wave speed is 250 meters per second, a decrease in wave speed to 20 meters per second will decrease the wavelength to about 32 kilometers, and the wave's height will grow approximately by the factor $\sqrt{400/32}$, or to about 12.5 meters (over 40 ft)! The tsunami does not actually gain any energy as it grows in this manner (this would violate our law of conservation of energy). What does happen is this: The energy that was originally spread out in a long, low hump on the ocean's surface becomes concentrated in a narrower and higher wall of water. The original wave that could transmit a modest force over a large horizontal distance transforms itself into a wave that pounds the shore with a tremendous force over a reduced distance. As we saw in Chapter 3, if a force exceeds a structure's ultimate strength, the structure is destroyed, regardless of the distance the force continues to move. Clearly, then, a tsunami's energy becomes much more hazardous to humans when it becomes concentrated into the higher waves of shorter wavelength that occur at coastlines. The total energy is unchanged, but the human consequences are amplified drastically.

To develop a complete mental picture of a tsunami, it's helpful to compare the wavelength to the depth of the water. In the open sea, a fairly typical tsunami may have a wavelength of around 400,000 meters (400 km, or 250 mi). In comparison, the Pacific Ocean has an average depth of only around 4,500 meters (4.5 km, or 2.8 mi). Viewed in comparison with a tsunami, then, the sea's average depth is only around 1% of the wavelength! As far as a tsunami is concerned, the world's oceans are but puddles, and the wave motion extends all the way to the sea bottom.

As a result, even in the very deepest oceans tsunamis propagate as shallow-water waves, with wave speeds in meters per second equal to 3.132 times the square root of the water depth in meters. This formula predicts that in water 4,500 meters deep, a tsunami will travel at the speed of 210 meters per second (470 mi/h)! In fact, such tremendous wave speeds have been corroborated by numerous observations. On May 22, 1960, for instance, an earthquake that measured 8.5 on the Richter Scale jolted a California-sized region of the seabed off the coast of Chile, causing local devastation and sending a tsunami racing out into the Pacific Ocean. Some 21 hours later and 17,000 kilometers distant, this tsunami swept into the Tohoku and Hokkaido districts of Japan with a wave height of up to 9 meters (nearly 30 ft), killing 180 people. Its average speed in crossing the entire Pacific Ocean was around 225 meters per second (503 mi/h).

In this behavior lies much of the horror of a tsunami, for it renders coastal cities vulnerable to events that may take place nearly halfway around the globe. Mid-Pacific islands such as the Hawaiian chain are doubly vulnerable, because they stand in the middle of a body of water that is ringed by seismically active coastlines, and because their coastal settlements are usually built with minimal tidal fluctuations in mind. On August 13, 1868, the island of Hawaii suffered severe damage from the Peruvian tsunami I discussed at the beginning of this chapter. On April 1, 1946, this island was again struck by a devastating tsunami (its height up to 9.1 m) spawned by an earthquake in the Aleutian Islands of Alaska. On May 23, 1960, a tsunami arriving from Chile swept into Hilo, Hawaii, with crests 5.8 meters in height.

The aerial view of Figure 5.7 shows the fourth crest of a tsunami climbing a Hawaiian beach in 1952. (This time the earthquake was in the Kamchatka Peninsula of the former Soviet Union.) From this vantage point the tsunami hardly appears threatening, and indeed on Mother Nature's grand scale it amounts to but a ripple. On a human scale, however, the devastation rendered by a tsunami can be catastrophic. The photograph in Figure 5.8 shows the first crest of the 1946 tsunami destroying a breakwater in Hilo.

Figure 5.7. On nature's grand scale, tsunamis are but ripples. Aerial view of the fourth crest of a 1952 tsunami climbing a beach on the northern shore of Oahu, Hawaii. (Photo courtesy U.S. National Oceanic and Atmospheric Administration.)

Figure 5.8. A tsunami striking Hilo, Hawaii, on April 1, 1946. (Photo courtesy National Geophysical Data Center.)

Contrary to what most people might expect, the leading edge of a tsunami need not be a wave crest. Around half the time a wave trough will arrive first, and observers will see the sea quickly recede to the horizon, as if King Neptune had pulled a giant plug in the ocean floor. This, in fact, may be the origin of the misleading term "tidal wave," for the uninformed could easily mistake such an event for an unscheduled, abrupt, and very low tide. When a trough does arrive first, it often has the unfortunate effect of attracting throngs of the curious to view the newly exposed sea bottom. This places them in the worst possible position to escape the first crest, which arrives perhaps 15 to 30 minutes later and bears down on them at freeway speeds. Needless to say, the geophysical data bases contain no shore-level photographs of what the imminent victims would see approaching.

For many years, an old hotel in the village of Captain Cook, Hawaii, had a sign prominently displayed in its lobby:

IN CASE OF TIDAL WAVE
1. Remain calm.
2. Pay your bill.
3. Run like hell.

The owner could afford to be glib because his hotel stood partway up a mountain slope, beyond the reach of any conceivable tsunami. Elsewhere on the island, the locals who recalled the losses of the 1960 event were inclined to take the threat more seriously.

When the tsunami warning sirens did go off in Hilo, at 8:30 AM on Sunday, May 22, 1960 (roughly 4 hours in advance of the first wave crest), some 37% of the male residents and 42% of the women evacuated. Some who did evacuate unfortunately grew restless and returned to their homes before the waves struck. This tsunami killed 61 and completely destroyed at least 500 dwellings.[8] Still, time dulls the collective memory, and it has now been thirty-six years since a significant tsunami has struck any of Hawaii's coasts. With no major event since 1960, one must wonder how today's Hawaiian population and its numerous tourists would respond to a tsunami evacuation order.

An average of three damaging tsunamis is generated somewhere in the world each year. There is no reason to believe that Hawaii, or for that matter any other vulnerable coast, has seen its last.

Pacific Coast of Nicaragua, 1992

Although the Pacific earthquake of September 1, 1992, registered a fairly significant 7.0 on the Richter scale, it was centered 100 kilometers (60 mi) off the Nicaraguan coast, and many on shore never even felt the ground tremors. It struck after dark, about 8 PM, and within the next hour a 300-kilometer (190 mi) stretch of coastline was inundated by tsunamis as high as 10 meters (33 ft). The waves killed 170, mostly sleeping children, and left more than 13,000 homeless. Thousands of buildings and boats were destroyed.

In the harbor of San Juan del Sur, two men relaxing in a boat were startled by a loud thump as their keel scraped bottom at a spot where the water was normally more than 6 meters deep. After successfully struggling to keep their boat from capsizing, they turned their attention toward shore. According to their later report, they saw the lights of the town through the back side of the wave crest that had just passed beneath them; then moments later, to their horror, the city suddenly went dark.

Elsewhere, numerous others reported observing extreme low water prior to the first wave crest. This was a tsunami whose first trough arrived before its first crest, a factor that probably contributed to the high survival rate. When the sea rolled out, some natives correctly interpreted the implication of the extreme low water and scampered to higher ground. In fact, many people who were outside at the time survived a direct strike by the tsunami.

Most of the dead were people who had been inside buildings when the wave hit.

To explain how so many managed to survive, we can draw upon our earlier theory of shallow-water waves and the particular fact that the wave speed is very sensitive to the depth of the water. If the trough of a tsunami arrives first, it is traveling upon water that is shallower than normal, and it has a lower wave speed (although a greater height) than if a crest strikes first. In most places along the shoreline, the Nicaraguan tsunami appears to have struck as a big but relatively slow-moving "slosh" rather than as a thundering breaker. This is consistent with the report that city lights were visible through the wave, for light does not penetrate the foam of a breaking wave. Hundreds, perhaps thousands, of people were buoyed up by this great hump of water and deposited hundreds of meters inland. Others reported spending a half hour or more clinging to floating debris in the harbor before managing to pull themselves ashore. Once again the implication is this: You stand a chance of surviving a tsunami if you're lucky enough to avoid any breaking portion of the wave. Shoreline structures, however, because they are anchored to the earth, will quickly be destroyed by the horizontal motion of even a relatively "gentle" tsunami.

Yet this particular event did raise a perplexing question for scientists. According to our current mathematical models, a magnitude 7.0 earthquake shouldn't pack enough energy to create much of a tsunami at all. There seems to be no evidence that an undersea landslide supplied the missing energy. We assume that the Nicaraguan event didn't violate the law of conservation of energy, because no exception to this principle has ever been noticed in many millions of observations of every imaginable type of phenomenon. Rather, we must assume that there is something wrong with either (1) our more detailed mathematical theory of tsunami formation, or (2) our measurement of the earthquake magnitude. This is hardly an esoteric point, for it strikes at the heart of our prospects for ever developing a comprehensive and reliable tsunami warning system. (Although a rudimentary warning system does exist at present, it protects only around 1% of the potentially vulnerable population around the Pacific rim.)

Has this ever happened before, a mild earthquake triggering a devastating tsunami? Yes, the catastrophic Japanese tsunami of 1896 was also preceded by only a relatively gentle ground tremor. The anomaly is rare, but it does occur. Several hypotheses have been proposed for the apparent discrepancy in energy between such "gentle" earthquakes and the occasionally disproportionate tsunamis they generate. The most likely explanation seems to be that some earthquakes release a good bit of their energy at periods that are longer than those our standard seismographic instruments are designed to respond to. The resulting seismic waves travel through the seabed as long, gentle undulations with periods of 20 seconds or greater. Neither humans nor their coastal structures will be affected by such long-period earthquake waves. The seas, however, are another matter. Raise or lower a few thousand square kilometers of seabed in just 20 seconds, and the sea surface will deform into a great tsunami that can easily catch coastal residents by surprise. At present, however, the scientific picture of this process is still incomplete, and perplexing questions remain that call for further research into the phenomenon of tsunami generation.

Galveston, Texas, 1900

On the evening of September 8, 1900, hurricane-driven storm waves drowned between 6,000 and 8,000 residents of the island city of Galveston, population 37,789. In terms of lives lost, this event continues to hold the record as the worst natural disaster in U.S. history. Some 3,600 houses and hundreds of other buildings were totally destroyed, and no human structure in the city completely escaped damage.[9] In nearby areas, the angry seas may have claimed up to an additional 4,000 lives.

Looking at a coastal map of the United States, one notices a string of long, narrow barrier islands extending down the entire eastern seaboard from New Jersey to southern Florida. Similar islands are found off the Gulf coast of south-central Florida, the stretch from the Florida panhandle to Louisiana, and along the complete length of the Gulf coast of Texas. Barrier islands are essentially very high sand bars that have been built up by wave action near coasts where the sea bottom slopes quite gradually. No barrier islands, or even beaches for that matter, are found along low-energy coastlines; the 210 kilometers (130 mi) of Florida's Gulf coast between Apalachee Bay and Crystal Bay, for instance, is lined only with heavy thickets of mangroves. Wherever one finds a barrier island, one is assured that the region is occasionally pounded by high-energy waves. One is also assured that, given the passage of enough time (say several centuries), every barrier island will be reshaped considerably as it creeps slowly toward the mainland. The lighthouse at Cape Hatteras, for instance, which in 1870 stood 460 meters from the water, by 1995 stood less than 60 meters from an often-angry surf. In 125 years, this particular barrier island beach receded the length of four football fields, while the opposite side of the island grew.

Early settlers were quick to recognize the instabilities of barrier islands, and few of them were foolhardy enough to build a permanent settlement on these shifting piles of sand. In 1838, however, a group of investors formed the Galveston City Company and began dividing up the real estate of Galveston Island, a barrier island near Houston, Texas. The venture was a huge success, and the new city thrived as a major shipping port. By 1900, a single five-block span of mansions boasted twenty-six millionaires.

At that time (this is no longer the case), the highest ground in the city of Galveston measured just 2.7 meters (8.7 ft) above mean sea level, or less than one story. From this point, the island tapered down to Galveston Bay on the northwest and the Gulf of Mexico on the southeast. Although the living quarters of many houses began at a height of only a meter or so above sea level, this seldom created a problem, because the tides themselves usually rose and fell only 0.6 meter or less. Occasionally, however, a tropical storm would raise the seas enough to flood the city. This happened three times in 1871 (stranding a schooner and three sloops in the city streets), again in 1875 (when the sea level rose 4 meters above normal and completely flooded the island), and in 1886, when there was a considerable loss of life on the mainland but only minor damage in Galveston. After the 1886 event, a commission considered building a seawall to protect the city but rejected the idea as too costly. Repairing minor damage every decade or so was expected to be much cheaper in the long run than incurring the cost of building and perpetually maintaining a seawall.

There is very little seismic activity in the Gulf of Mexico, and no tsunami has occurred in this region in historical times. On the other hand, tropical storms and hurricanes are quite common in the Gulf. The low barometric pressure associated with such storms often raises the sea level to a considerable height over a very wide area, and wind-driven waves riding on top of such a "storm surge" or "storm swell" can be every bit as devastating as a great tsunami. The effect is particularly disastrous when the storm passes through slowly rather than quickly.

This was the case with the hurricane of September 8, 1900. We don't really know the peak wind speed, because the Weather Bureau's wind gauge blew away when the winds hit 84 miles per hour at 5:15 PM, and the storm continued well into the night. Judging from the nature of the wind damage, however, it seems unlikely that this hurricane ever produced sustained winds much in excess of 100 miles per hour. Virtually all of the devastation was caused by sea waves rather than wind.

The unnamed hurricane's arrival was no surprise, for the local meteorologist had been receiving regular telegrams over the previous few days giving updates on the storm's progress through the Atlantic, from its glancing blow on southern Florida to its entrance into the Gulf of Mexico. The previous afternoon, he'd already noted minor flooding on the lowest parts of the island, despite the brisk offshore wind from the north (a condition that normally reduces the tides rather than exaggerating them). City residents, however, were not particularly alarmed, and no one felt it necessary to evacuate the island. On the morning of September 8, with the barometer dropping and a heavy rain falling, most workers still went to their jobs, and many women and children went to the beach to watch the pounding breakers. By early afternoon, when the wind speed first broke into the hurricane category (119 km/h or 74 mi/h), all shoreline structures had already been demolished by the waves, and a huge wall of debris was being driven farther into the city by each successive impact of the relentless breakers. It was now too late to evacuate, for no boat or barge had any chance of surviving a transit to the mainland.

By 6 PM, the sea was rising three-quarters of a meter (2.6 ft) per hour, and the wind was beginning to shift to the east. At 7:30 PM, in one great surge the sea level jumped 1.20 meters (4 ft) in just 4 seconds. The eye of the hurricane apparently passed just west of the island sometime between 8 and 9 PM. By this time, the entire island was under water at least 3 meters (10 ft) deep, and many of the waves rose 6 or 7 meters higher. For several terrible hours Galveston Island ceased to exist, and the fate of the living depended on the durability of those taller buildings whose upper stories poked above the frothing sea. The waves ripped up long sections of streetcar tracks, still lashed together with ties, and battered them into rows of houses that quickly broke into splinters. A giant wall of debris, several stories high and roughly parallel to the beach, ultimately anchored itself about six blocks inland (Fig. 5.9). From here to the original shoreline, everything was scoured clean by the waves. Serendipitously, the great wall of wreckage acted as a breakwater that for several hours protected the rest of the city from total destruction. Around 1:45 AM, the sea began to subside.

The next morning, survivors found one-third of the island scraped clean and the remainder battered almost beyond recognition. Everything was covered by a thick, foul-smelling slime. One observer counted forty-eight bodies dangling in the trusswork of a partially demolished railroad trestle. Initial attempts were made to bury the thousands of corpses at sea, but when

Figure 5.9. Wreckage of Galveston, Texas, after the hurricane of September 8, 1900. This wall of debris prevented the storm-driven waves from penetrating farther inland. (Photo courtesy Rosenberg Library, Galveston.)

the bodies began to wash back up on the beach it became necessary to stack them on the mountains of debris and burn them. From one end of the island to the other, these funeral pyres burned continually for several days and nights.

The horrors of the disaster drove many survivors from the island, never to return. Property values, for those structures that remained standing, plummeted to ten cents on a dollar. The next year, however, when it was apparent that Galveston was indeed going to rebuild, a board of engineers was commissioned to do an in-depth study of the disaster and to make a recommendation on how such a catastrophe might be prevented in the future. In January of 1902, these engineers delivered their report, which included the audacious recommendation that the entire city should be raised in elevation as much as 3.4 meters (11 ft) and walled off from the sea.

What followed was one of the most amazing engineering feats of the early twentieth century.[10] To carry in heavy machinery, materials, and fill dirt, temporary railroads and canals were constructed throughout the length and width of the city. Large masonry structures like St. Patrick's Church, which weighed some 3,000 tons, were lifted on hundreds of hydraulic jacks as new foundations were built beneath them. Almost three thousand buildings were raised in this manner. At the same time, it was necessary to relocate water and sewer lines, electric lines, streets, sidewalks, trees, and gardens, all without making it impossible for residents to go about their daily business.

A concrete seawall was built along the Gulf, 4.9 meters (16 ft) wide at its base, 1.5 meters (5 ft) wide at the top, and extending to a height of 5.2 meters (17 ft) above mean low tide, its seaward face concave, so that it would deflect waves upward rather than allowing them to bear against it with their full force. On the seaward side of the wall, an 8.2-meter (27-ft) apron of granite riprap was laid over the beach to further sap the energy from any large waves. The city side (Fig. 5.10) was backfilled with sand to give it a gentle slope down to the level of the top of the wall. The initial seawall was 5 kilometers (about 3 mi) long; it was later extended, so that it now runs a total length of 16.15 kilometers (10.04 mi.). Raising the city and completing the initial section of seawall took almost seven years.

This engineering project has demonstrated its effectiveness in several

Figure 5.10. Construction of the Galveston seawall, 1902–10. Behind the wall, the entire city was raised as much as 3.4 meters (11 ft). (Photo courtesy Rosenberg Library, Galveston.)

hurricanes, the first of these in 1909, before the wall was even completed. Today, beyond the western end of the wall, even a casual observer will notice that the unprotected beach has eroded inland about 50 meters (160 ft). In front of the wall, however, there is no longer any beach at all: only riprap.

Clearly, it is not economically feasible to protect every inhabited barrier island with an engineering project of the scope of Galveston's, nor would the residents of recreational beachside communities be pleased with the result: destroying the beach to save the homes. To build on a barrier island is to assume a risk.[11] The waves will come, the beaches will creep, and flooding from the sea will someday innundate all coastal structures we build in such places. The gamble homebuilders take is that damaging sea waves are not likely to strike soon. But then again, they just might. Our modern science remains woefully inadequate when we use it to try to predict what Mother Nature might have up her sleeve during our lifetimes.

Notes

1   Most of my account here is drawn from L. G. Billings, Some personal experiences with earthquakes, *National Geographic,* Jan. 1915, 57–67. The U.S. Navy also has records and photographs relating to the event, and shorter articles can be found in numerous other sources, including those cited in note 4 to this chapter.

2   Two days later, August 15, another major earthquake struck to the north in Peru and Ecuador and may have claimed as many as 40,000 lives. Some sources list only this second event, which apparently did not generate a major tsunami. Billings, writing forty-seven years after the event, also seems to have gotten the date of his own adventure wrong; he gives it as August 8, whereas U.S. Coast and Geodetic Survey sources give it as August 13. Readers who wish to explore the literature further on this incident should be aware that there were two separate geophysical events, and that there may be discrepancies in the literature regarding the dates of one or both.

3   A more formal definition is this: A wave is a disturbance in which (1) at every set of spatial coordinates the displacement of the medium is a function of time, and (2) at every instant in time the displacement of the medium is a function of its spatial coordinates.

4   W. Bascom, *Waves and beaches* (New York: Anchor, 1980).

5   The numerical coefficients in this and the preceding equation are not arbitrary but are related to the gravitational acceleration, g, which in international

units has the standard value 9.80665 m/s^2. The value 3.132 is $\sqrt{g}$, and 1.249 is $\sqrt{(g/2\pi)}$.

6 K. Lida, Magnitude, energy, and generation mechanics of tsunamis; Lida, *On the estimation of tsunami energy,* International Union of Geodesy and Geophysics Monograph no. 24 (Toulouse; France: IUGG, 1963), 7–18 and 167–73.

7 E. R. Scidmore, The recent earthquake wave on the coast of Japan, *National Geographic,* Sept. 1896, 285–9.

8 R. Lachman, M. Tatsouka, & W. J. Bonk, Human behaviours during the tsunami of May, 1960, *Science,* 133 (1961), 1405–9.

9 W. J. McGee, The lessons of Galveston, *National Geographic,* Oct. 1900, 377–83. Reasonably credible accounts of the disaster can also be found in many other publications, among them G. Cartwright, The big blow, *Texas Monthly,* Aug. 1990, 76–87.

10 A more detailed description of the engineering feat can be found in D. Walden, Raising Galveston, *American Heritage of Invention and Technology,* 5(3) Winter 1990, 8–18.

11 For a further discussion of the effects of overdevelopment on barrier islands, see S. Kemper, This beach boy sings a song developers don't want to hear, *Smithsonian,* Oct. 1992, 72–85. Also, K. Wallace & O. H. Pilkey, Jr., *The beaches are moving: The drowning of America's shoreline* (Durham, N.C.: Duke University Press, 1983).

6

Earth in Upheaval

Alaska, 1964 and Before

At 5:36 PM on March 27, 1964, a tremendous burst of seismic energy jolted southern Alaska and permanently resculptured 200,000 square kilometers (78,000 sq. mi) of land surface, a region much larger than the entire state of Florida. The surface elevation rose as much as 10 meters in some places, dropped 2 meters elsewhere, and huge fissures and cracks opened up throughout the region. At the same time, underground forces thrust thousands of square kilometers of sea floor above sea level, stranding barnacles and other marine creatures high and dry and sending devastating tsunamis racing toward Hawaii and the coast of California. In Alaska, the city of Anchorage was particularly hard hit (Fig. 6.1), incurring hundreds of millions of dollars in property damage, even though the earthquake's epicenter lay 130 kilometers (80 mi) to the east. Much of the city's structural damage resulted from a sudden liquefaction of the soil beneath the foundations of buildings, and some houses that stood on slopes slid as far as 300 meters (1,000 ft) from their original sites.

The official death toll of 131 belies the enormous amount of energy released in this earthquake that measured 8.6 on the Richter Scale; in fact, the event resulted in the greatest total mass of vertical movement in a century of scientific earthquake measurement.[1] Fortunately, most of the affected area was sparsely settled, and most of the structures were wood. As we saw in Chapter 3, wood-framed buildings are quite resilient in the face of dynamic loading, and they generally survived the 1964 event, as long as their foundations remained intact.

Although thousands of aftershocks were recorded at seismographic observatories in the following months, none of these events was very serious. The Alaskan crust settled down, and there has not been another violent tremor in the last thirty years. Yet there are compelling reasons, both theoretical and historical, to believe that southwestern Alaska and the Aleutian

Islands will continue to experience major earthquakes from time to time indefinitely into the future.

Written records of Alaskan earthquakes begin with Vitus Bering's 1737 expedition, when his astonished chronicler also described a large tsunami. The Russians continued to document ground tremors over the following century or so, and they noted a particularly violent series of earthquakes and volcanic eruptions in the period 1820–5. The United States purchased Alaska from Russia in 1868, when this huge territory had an official population of just 32,996 Native Americans and 430 others. These numbers grew very slowly until the discovery of gold in 1898 attracted an influx of 30,000 people in that year alone. Among the numerous unpleasant surprises that greeted these new immigrants was the phenomenon of the earthquake. Survivors of the great 1899 earthquake reported that the ground shook for a full three minutes; in that same event, a part of the coast of Disenchantment Bay was permanently raised by 14.4 meters (47.3 ft), which set a still-standing record for such a displacement within historical times. Although there were relatively few humans around to be affected by that great upheaval of 1899,

Figure 6.1. A school in Anchorage, Alaska, after the Richter magnitude 8.6 earthquake of March 27, 1964. (Photo courtesy National Geophysical Data Center.)

nearly 200,000 people were directly affected by the magnitude 8.6 earthquake of 1964. Today, an earthquake of the same intensity would be even more disastrous in its human impact.

Seismographic recordings are generally available after about 1903, and this allows scientists to rate and compare the magnitudes of earthquakes on a reasonably consistent numerical scale. Major earthquakes occurred in the Aleutian Islands in 1903 (Richter magnitude 8.3) and 1906 (magnitude 8.3), in the Bering Sea in 1938 (magnitude 8.7), and in southern Alaska in 1949 (magnitude 8.1).[2] History alone, then, suggests that there will probably be future events of this type in the same general geographical region. To proceed beyond this historical pattern, however, we need to explore the underlying geophysical mechanisms that give rise to earthquakes. Let me begin with an overview of what scientists have learned (so far) about detecting and measuring earthquakes.

A Brief History of Seismology

When a city that may have taken centuries to build is shaken to the ground in just seconds, contemporary chroniclers tend to be distracted from their usual preoccupation with human power at least long enough to document that the event took place. Such accounts, although often fragmentary, date back to the earliest Greek and Roman historians. By the late 1600s, a few scholars sifting through these old sources began to compile lists of the documented earthquakes; the earliest seems to be Vincenzo Magnati's 1688 list of ninety-one major earthquakes that occurred in the period A.D. 34 to A.D. 1687. Over the next two centuries, a dozen or so others published their own lists, often explicitly restricted to a particular geographical area or a particular period in time (e.g., one chronicles 1,186 shocks in Italy for the period 1783–6). To the extent that these lists overlap, they are often contradictory in relevant details. An equally serious shortcoming is that their entries reflect contemporary population distributions, their geographical accessibility, and the psychology of mass hysteria more than they describe anything approaching an objectified geophysical data base.

With the invention of the telegraph in 1840, it became possible to communicate reports of earthquakes much more efficiently, and information (along with misinformation) mushroomed. Alexis Perry catalogued more than 21,000 earthquakes for the years 1843–71; Robert Mallet (more discriminating in his criteria) described 6,831 events for the period 1606 B.C. to A.D. 1850; Guiseppe Mercalli (1883) listed more than 5,000 earthquakes from 1450 B.C. to A.D. 1881 in Italy alone; Carl Fuchs (1886) developed a monumental list containing nearly 10,000 entries; and John Milne (1895) described 8,331 earthquakes recorded just in Japan. Jean Baptiste Bernard, however, seems to hold the one-man endurance record for this type of research; working for twenty-one years on the project, by 1906 he'd accumulated a list of earthquakes from throughout the world that included 171,434 entries!

Few, if any, of these early lists were of permanent significance.[3] Through their inconsistencies, however, such lists did make it clear that a uniform scale for describing earthquake intensity was desperately needed. In 1883, Guiseppe Mercalli (himself one of the list makers) rose to the occasion and proposed the Mercalli scale, a system still based on somewhat subjective observational descriptions. Others adopted it but, before long, began to fiddle with its criteria. The scale was officially modified in 1912, then once again in 1931; in the latter form it is still sometimes used today. An abbreviated version of the Modified Mercalli Intensity Scale is presented in Table 6.1.

The main continuing appeal of the Mercalli scale is that it does not depend at all on the use of scientific instruments but on ordinary human observations. Anyone who can make and assess the required observations can assign a Mercalli intensity to an earthquake. At the higher intensities, one need not even experience the event firsthand to assign the number, for the relevant criteria can be established through an examination of the damage left behind.

The Mercalli scale does, however, have its shortcomings: (1) it applies only to populated areas (a fact that becomes obvious as soon as you read the criteria); (2) it does not allow for fractional intensities (in fact, Roman numerals are used, so no one is likely to be tempted on this issue); and (3) it does not give any indication of the strength of the *source* of the earthquake (a low Mercalli intensity does not distinguish between a mild earthquake nearby and a strong one a greater distance away).

Nevertheless, when Mercalli numbers began to be applied to earthquakes, patterns started to emerge from the hodgepodge of lists. Well before 1920, it became clear that the most seismically unstable regions of the earth are associated with surface features where the earth's crust is most severely corrugated – for example, mountains and rifts (whether above or below the sea). Further, there are two broad bands on the globe that together account for more than 90% of the significant earthquakes: one of these bands circles the Pacific Ocean; the other extends in a shallower arc from Indonesia through the Himalayas to the Mediterranean.

Table 6.1. *Modified Mercalli Scale for earthquake intensities, abbreviated summary.*

Intensity	Description
I.	Not felt by humans.
II.	Felt by persons at rest on upper floors.
III.	Hanging objects swing. May not be recognized as an earthquake.
IV.	Standing cars rock. Windows, dishes, and doors rattle. Wood-framed walls may creak.
V.	Felt outdoors. Sleepers awakened. Liquids disturbed, pictures moved, doors set swinging.
VI.	Felt by all. Walking difficult. Windows and glassware broken. Furniture moved or overturned. Weak plaster and masonry cracked. Church and schoolbells may ring.
VII.	Difficult to stand. Noticed by drivers. Furniture broken. Weak chimneys and unreinforced brickwork may fall. Waves on ponds.
VIII.	Steering of cars affected. Partial collapse of unreinforced masonry structures. Frame houses may move off foundations. Changes in flow of wells and springs. Cracks in wet ground and on steep slopes.
IX.	General panic. Unreinforced masonry destroyed, reinforced masonry damaged. Underground pipes broken, serious damage to reservoirs. Conspicuous cracks in ground.
X.	Most masonry and frame structures destroyed. Damage to dikes and dams. Large landslides. Some rails may be bent.
XI.	Rails seriously bent. Underground pipelines completely out of service.
XII.	Damage nearly total. Objects thrown into air. Large rock masses displaced.

Note: For the original 1931 version of the scale, see H. O. Wood & F. Neumann. Modified Mercalli Intensity Scale of 1931, *Seismological Society of American Bulletin*, 53 (5), 979–87. For the 1956 version, see C.F. Richter, *Elementary seismology* (San Francisco: Freeman, 1958), 137–8.

Scientists, meanwhile, looked for a more "scientific" way to measure earthquake strength – one linked to the recordings of an unbiased instrument. To design any instrument, however, you first need to have some quantitative understanding of the phenomenon you are trying to measure. This is obviously a Catch-22, for unless you already know something about the relevant characteristics of the phenomenon, you cannot build an instrument that will tell you about the very characteristics you need to know about. Given that earthquakes are sporadic and unpredictable by their very nature, progress in designing seismographs proceeded slowly.

The first device specifically designed to record earthquakes was apparently built in China in A.D. 132. This was a circle of eight sculptured bronze dragons, each holding a metal ball in its mouth, and, directly below, a corresponding circle of open-mouthed bronze toads. A strong earthquake would make a dragon drop a ball into the mouth of a toad, and the particular toad was expected to indicate the direction of the earthquake source (an incorrect assumption, it turns out). This device was a beautiful work of art but of dubious value as a scientific instrument, for any earthquake strong enough to be registered in this manner would already be quite apparent to everyone, and the instrument was incapable of supplying any additional information about such an event.

By the early 1700s, it had become common knowledge that strong earthquakes disturb the water surfaces of ponds and lakes, and this phenomenon was exploited in several early seismoscopes. Most of these devices used some variation on a vessel of liquid mercury that would spill, or at least slosh around, leaving a record of its motion. None of these devices was sensitive enough to be of much scientific value.

A more fruitful approach was to use a pendulum. It had long been noticed that bells in churchtowers often rang spontaneously during a strong earthquake and that pendulum clocks often stopped. Beginning in 1841, James D. Forbes experimented with various pendulum arrangements, and he eventually built a "seismoscope" consisting of a pencil attached to an inverted pendulum, which successfully recorded two earthquakes. Unfortunately, it failed to respond to most of the several dozen other earthquakes that were felt in the area where it was set up.[4] Meanwhile, geophysicists who were trying to measure the subtle effects of the Sun's and Moon's gravity on Earth were making considerable progress with instruments employing a heavier pendulum, and they were often finding (to their annoyance) that such instruments would go into spasms of uncontrolled jiggling during minor earthquakes. Closer examination revealed that the heavy pendulum itself wasn't jiggling at all; rather, the instruments were recording the vibrations of the ground relative to the pendulum, which, because of its inertia, remained pretty much at rest.

In Italy in 1875, Filippo Cecchi put these ideas together and built the first successful seismograph. The device used two heavy pendulums, suspended in such a way that one detected north–south motion and the other detected east–west motion (orientations still used today). At the same time, a third mass suspended on a spring permitted a measurement of the vertical component of the earthquake motion. Over the next few years, John Milne (working in Tokyo) made considerable improvements in the sensitivity of this instrument. Useful seismographic recordings of ground motion date from the Japanese earthquake of November 3, 1880. By the time of the 1906 San Francisco earthquake, described in Chapter 3, scientists were able to compare seismograms of ground motion that had been recorded simultaneously at a number of observatories in different parts of the world.

There have been many improvements since, both in terms of sensitivity and in the method of recording data; computer printouts, for instance, have generally replaced the earlier strip-chart recordings. Many modern seismographs no longer measure the motion of the earth relative to a suspended inertial mass; instead they use electronic sensors to measure the strain deformation of the earth directly, usually between two points in a long underground tunnel. In this manner, it is possible to measure crustal movements as small as 0.001 millimeters over lengths of around 25 meters. Problems, nevertheless, remain. Even today, it is difficult to make reliable measurements of very long-period earthquake waves (30-s periods or greater). Moreover, a very strong earthquake will saturate the most sensitive instruments, in the same manner as if you tried to weigh a car on a bathroom scale. As a result, seismographic observatories need to keep a whole array of instruments in continuous operation, some for weak motions and others for strong motions.

The Mercalli Intensity Scale, in one of its three principal versions, was used almost universally for some fifty years. With the progress in instrumentation, however, prospects improved for linking earthquake size to actual seismographic recordings of ground motion. By 1930, it was possible to combine seismographic data from different observatories to pinpoint the geological sources of most earthquakes. What remained was to develop an objective measure of the absolute *magnitude,* or source strength, of an earthquake.

In 1935, Charles F. Richter developed an enthusiastically accepted procedure and numerical scale for assigning earthquake magnitudes on his Richter Scale. An earthquake's Richter magnitude is determined by reading the maximum ground motion recorded by a seismograph, adjusting this value to reflect a "standard" distance from the source (100 km), correcting for any peculiar characteristics of the particular instrument used, then using a mathematical formula to relate the result to a logarithmic numerical scale. Figure 6.2 shows the basic relationship in graphical form. The Richter Scale has no top or bottom but can generally be considered to run from 0 to 9. Each increase of 1 on this scale represents a factor of 10 times the ground-motion amplitude, and an increase of 2 represents a factor of 10×10, or 100. For example, at a distance of 100 kilometers from the source, a magnitude 8.3 earthquake generates 10 times the shaking amplitude of a magnitude 7.3 earthquake. Similarly, a magnitude 5.6 earthquake shakes the ground with only 1/100 the amplitude of a magnitude 7.6 event. Although these comparisons technically apply only at the standard 100-kilometer distance from

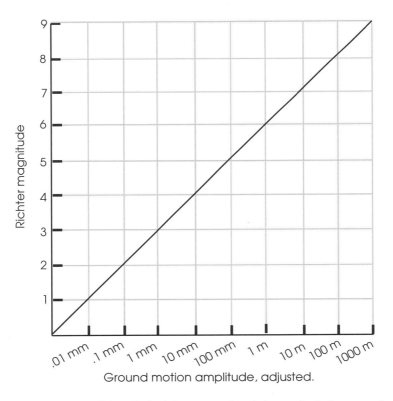

Figure 6.2. The Richter Scale defines an earthquake's magnitude in terms of the seismographically recorded ground motion, scaled to a standard 100-kilometer distance from the source.

the source, until quite recently they were usually treated as a measure of the energy released at the source itself.

Earthquakes, it turns out, can differ greatly in many respects: The physical source can be deep or shallow; the source can be a large slip in a concentrated region of a fault or a smaller slip over a more extended region; the source might release greater portions of its energy in shorter- or longer-period waves; and so on. For this reason, it soon became apparent that a single Richter Scale did not give an honest comparison of the energy released in all types of earthquakes. Although it remains common practice to use the *Richter magnitude* (designated M_L) for moderate-sized local earthquakes, a better measure for larger earthquakes seems to be the *moment magnitude* (M_W), which involves a series of different seismographic measurements and a somewhat more involved calculation. These two scales often disagree; the 1964 Alaskan earthquake, for instance, had magnitude $M_L = 8.6$ but $M_W = 9.2$. Moreover, different seismographic observatories often report slightly different M_L or M_W scores for the same earthquake. It now appears that complete consistency in assigning earthquake magnitudes is probably an unrealistic goal, for when the natural phenomenon itself is inherently fuzzy and irreproducible, no amount of mathematical fiddling can be expected to force all of the data into agreement.

While these developments in earthquake measurement were taking place, seismologists also made considerable progress in mapping the interior of the earth by analyzing how seismic waves travel between distant parts of the globe. This research helped establish the theory of plate tectonics, discussed in the next section, which was on a fairly firm footing by the early 1960s and explained why earthquakes are more common in some regions than others. About the same time, researchers began to propose more detailed theories of the mechanisms that produce earthquakes. In 1962, Japanese seismologists adopted earthquake prediction as a formal goal, and in the United States, in 1977 the Earthquake Hazards Reduction Act established prediction as a formal objective of U.S. government-sponsored seismological research. Although to date progress toward this goal of predicting the time, location, and size of earthquakes has been disappointing at best, seismology is still a very young science, and only time will tell where continuing seismological research will lead.

The Mechanism of Earthquakes

Earth's radius measures 6,378 kilometers (3,964 mi), but its solid crust extends down only 25 to 60 kilometers below the continents and just 4 to 8 kilometers below the deep oceans. In other words, the terra firma on which we humans spend our lives accounts for less than 1% of Earth's radius. What is below this thin crustal shell? A region called the *mantle* extends from the base of the crust to a depth of around 2,885 kilometers. The mantle has a complex internal structure in places, particularly below the most seismically active regions of Earth's surface. Over short periods of time most of Earth's mantle behaves as a solid, but over geological time spans it flows slowly, like a giant gob of putty. Deeper still is Earth's core, at least the top 2,200 kilometers of which is liquid.

Because of this viscous interior, the great land masses of Earth do not stay put forever. Earth's crust is not a single unified structure; it is, rather, a collection of well-defined crustal plates that grind past each other, under and over each other, and in places recede from each other. Figure 6.3 shows Earth's principal crustal plates. It appears to be far from an accident that most earthquakes and volcanoes occur near the boundaries between these plates. The theory that seeks to explain seismic and volcanic activity on this basis is referred to as the *theory of plate tectonics.*

In looking at Figure 6.3, we immediately notice that the plate boundaries do not correspond neatly with the outlines of the continents. North of India, for instance, is the inland boundary between the northeastward-moving Indo-Australian Plate and the southeastward-moving Eurasian Plate; here we find that great upswelling of Earth called the Himalayan Mountains, and, just to the north in China, a region that is subject to frequent and devastating earthquakes.[5] Along the western coast of South America, we see another collision boundary between plates that have opposing motions; here the crush has resulted in the Andes Mountains, whose continued growth is accompanied by numerous earthquakes and volcanoes. Meanwhile, the northwestward motion of the Pacific Plate acts as a giant wedge that lifts southern Alaska and the Aleutian Islands, often abruptly and catastrophically – as happened in 1964.

Boundaries between plates are often visible on the ground surface. In the photograph in Figure 6.4, for instance, we see a line of trees in Guatamala that was suddenly shifted horizontally some 3.25 meters (10.7 ft) during the earthquake of February 4, 1976. This earthquake killed 23,000, injured some 76,000, and resulted in at least $1.1 billion in property damage. The line of slippage shown in the photograph is actually a portion of the Motagua Fault, which defines the boundary between the North American Plate and the Caribbean Plate.

The Motagua Fault is an example of a *strike-slip fault,* where the plate on one side of the boundary moves horizontally relative to the plate on the other

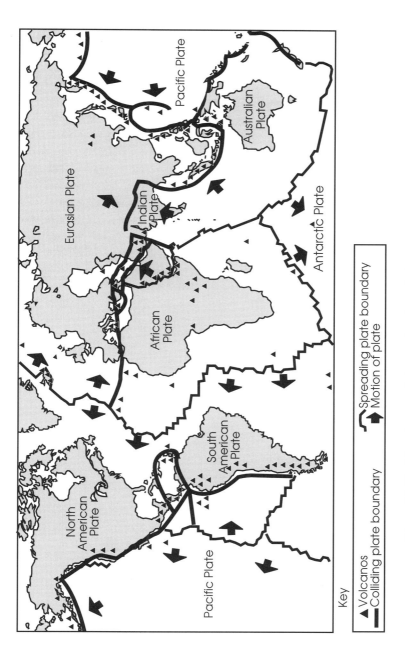

Figure 6.3. Earth's major crustal plates and regions of seismic and volcanic activity.

Key

▲ Volcanos
━ Colliding plate boundary
Spreading plate boundary
Motion of plate

Figure 6.4. A horizontal offset of 3.25 meters occurred on the Motagua Fault in Guatemala during the Richter magnitude 7.5 earthquake of February 4, 1976. The stake at the right marks the original position of the line of trees at the left. This earthquake claimed 23,000 lives. (Photo courtesy National Geophysical Data Center.)

side. If, on the other hand, one plate is thrust over or under a second plate, the boundary is referred to as a *dip-slip fault,* and the result is sometimes visible as a vertical escarpment on the surface. It is also possible for two plates to pull apart from each other; in this case, the boundary manifests at the surface as either a rift valley (usually below the ocean) or a region of rift volcanic activity (as in Iceland).

Of course, geoscientists make other, finer classifications, and none are Aristotelean in the sense of being either–or. Strike-slip faults are often combined with dip-slip motion, and rifts may be associated with strike-slip motion. Although we humans find such classifications useful in trying to make sense of nature's bewildering complexity, we should never delude ourselves into thinking that the classifications *are* the reality. This said, I'll spare the reader an excessive digression into the nomenclature of faults.

The important point is this: There are places where intersecting crustal plates creep smoothly past each other at the fault lines, and there are other places (even on the same fault) where the plates lock up for decades or even centuries while internal strain builds up. A winery actually straddles the the

San Andreas Fault near Hollister, California. Cracks in this structure and an adjoining culvert steadily open up at an average rate of about 1.5 centimeters (1 in.) per year. Meanwhile, other sections of the San Andreas Fault have been locked up for decades. Indirect evidence suggests that there are other faults in California that haven't slipped in several centuries – so long, in fact, that the natural forces of erosion have long obliterated any surface evidence that the faults even exist. There is evidence that one such long-locked fault may pass directly beneath central Los Angeles.

Theoretically, if the earth's crustal plates all slid smoothly against each other, there would be no earthquakes associated with geological faults (although a few might still be connected with volcanoes). Most earthquakes result from a frictional binding effect that prevents the crust from moving smoothly along a segment of a fault. When the internal stresses in a fault grow to the point where they exceed the forces that lock it up, the crust suddenly leaps into motion, like a broken clock spring. The result is an earthquake.

Earthquakes, then, seem to be Mother Nature's way of relieving tectonic stress. The bigger the quake, the greater the stress that is relieved, and the longer it ought to take for the stress to build up again to the same level. This is the essence of the *seismic gap theory*, which states that (1) strong earthquakes are unlikely in regions where weak earthquakes are common, and (2) the longer the quiescent period between earthquakes, the stronger the earthquake will be when it finally does break loose.

For the past few decades, valiant efforts have been made to translate the seismic gap theory into a mathematical tool for forecasting earthquakes. So far the statistics seem to work, at least approximately, for the relatively rare earthquakes of magnitude 8 and greater. (Valparaiso, Chile, for instance, seems to get a magnitude 8 earthquake roughly every eighty to ninety years.) Unfortunately, to date, the seismic gap theory has been grossly inconsistent in making even statistical predictions of earthquakes of lower magnitude, which are those that cumulatively cause most of the damage worldwide. Table 6.2 summarizes the relative frequency of earthquakes of different Richter magnitudes. A magnitude 7 earthquake can cause great property damage and even a significant loss of life yet still fail to relieve enough of Earth's internal stresses to render a follow-up quake of equal magnitude unlikely. Even magnitudes as low as 6 on the Richter Scale are sometimes devastating (Table 6.3). From a human perspective, then, there is a considerable interest in learning more about the more frequent and cumulatively more damaging lower-magnitude quakes.

One complication is that the boundaries between Earth's crustal plates are often fractured into a vast network of minor faults that intersect the

Table 6.2. *Approximate number of earthquakes per year, worldwide, exceeding given Richter magnitudes*

Richter magnitude	Number exceeding this magnitude
8	2–3
7	20
6	100
5	3,000
4	15,000
3	100,000

major fault lines. When an earthquake relieves the stress in any one of these faults, it may pile additional stress on another fault in the network. In this way (and contrary to the seismic gap theory), a series of small earthquakes can sometimes actually *increase* the probability that a large quake will follow in the same general area. Further complicating matters is the fact that some faults run deep into Earth's mantle, where the tremendous pressure makes them more prone to lock up. Near the surface, the walls of a fault may be gliding over each other relatively smoothly, while deep below the stress is building up to a possibly catastrophic event. It remains beyond the capability of our modern instrumentation to probe the depths of Earth directly to see what is going on. All we can do at present is to analyze the waves that travel through Earth when an earthquake does occur, then work backward in time to develop plausible hypotheses about how the event may have been generated. We are all awash in a sea of seismic ignorance, particularly with regard to the origin and escalating prospects of the milder tremors. This should not be construed as a cynical view on my part. To the contrary, where humankind's pockets of ignorance have run deepest, science has historically progressed the fastest. It is the *direction* of future scientific progress that remains the open question, and one that puzzles every public administrator faced with the decision of how to allocate limited research funds among hundreds of eager and underfunded scientists.

Primary (P-) and Secondary (S-) Waves

One of the most horrifying aspects of earthquakes is that they often do great damage at considerable distances from the source. Nature abhors concen-

Table 6.3. *Some devastating earthquakes with Richter magnitudes less than 7*

Date	Place	Richter magnitude	No. of Deaths
1960, Feb. 29	Agadir, Morocco	5.9	14,000
1963, July 26	Skopje, Yugoslavia	6.0	1,100
1966, Aug. 19	Eastern Turkey	6.9	2,520
1967, July 29	Caracas, Venezuela	6.5	250
1971, Feb. 9	San Fernando, Calif., USA	6.5	65
1972, Apr. 10	Southern Iran	6.9	5,057
1972, Dec. 23	Managua, Nicaragua	6.2	10,000
1974, Dec. 28	Pakistan	6.3	5,200
1975, Sept. 6	Lice, Turkey	6.8	2,312
1976, May 6	Friuli, Italy	6.5	965
1982, Dec. 13	Northern Yemen	6.0	2,800
1983, Mar. 31	Southern Colombia	5.5	250
1986, Oct. 10	El Salvador	5.4	1,000
1988, Aug. 20	India/Nepal	6.5	1,000
1988, Dec. 7	Northwestern Armenia	6.8	55,000
1989, Oct. 17	San Francisco, Calif., USA	6.9	62
1990, May 30	Northern Peru	6.3	115
1991, Feb. 1	Pakistan/ Afghanistan	6.8	1,200
1992, Mar. 13, 15	Eastern Turkey	6.2 and 6.0	4,000
1992, Oct. 12	Cairo, Egypt	5.9	450
1995, Jan. 17	Kobe, Japan	6.9	5,500

trations of kinetic energy in an extended body like Earth, and when a geological fault suddenly slips, the kinetic energy it releases spreads outward from the source as a series of spherical wave fronts. Within the planet, these seismic waves bend and reflect at the interfaces between the crust and the mantle, and between the mantle and the outer core. At the upper surface of the crust, they transfer some of their energy to the seas (the origin of tsunamis) and the atmosphere (the origin of the "rumbling" sound often associated with earthquakes). The remaining energy radiates outward along

Earth's crust like a series of ripples in a pond. There are numerous eyewitness accounts of the ground surface "wiggling like a snake" during strong earthquakes.

In talking about seismic waves in Earth, I use the word "Earth" in reference to our planet, which is in fact mostly rock at the depths where earthquakes originate. In most regions, the shallow layer of topsoil has little effect on seismic waves. But where the topsoil is deep, and particularly where it is deep and wet, earthquake waves slow down and exhibit the same behavior as a tsunami that enters shallow water: The waves grow in amplitude to destructive proportions. An earthquake wave that enters a region of clay or fill dirt is like a storm-driven sea wave crashing on a beachfront. This effect explains why a single 25-square-kilometer section of Mexico City suffered the brunt of the damage in the 1985 earthquake whose epicenter was a full 350 kilometers distant. The dirt that had been dumped there to fill an ancient lake bed in effect offered a "shore" for the seismic waves to crash upon.

With the development of reliable and sensitive seismographs in the late 1800s, it soon became apparent that Earth propagates two distinct types of seismic waves, which travel at different wave speeds. The fastest, and therefore the first to arrive at any seismological observatory, is the *P-wave* (*P* for *primary wave*, or *pressure wave*, or "*push*" wave). Next comes the *S-wave* (*S* for *secondary wave*, or *shear wave*, or "*shake*" *wave*), which is considerably more destructive to human-built structures. The P-wave is identical to a sound wave, and if its period is shorter than about 0.05 seconds, it will actually be heard as a low rumbling sound coming from the earth. If a P-wave's period is longer than 0.05 seconds (the usual case), the ear will not pick it up, but it will still be felt as a trembling sensation in the body's internal organs. A P-wave may kick up dust from Earth's surface, but by itself it will seldom cause any serious lateral shaking in buildings. The S-wave that follows, however, is what often shakes buildings to pieces. Roads turn into snaking ribbons, and bridges sway like swings. The diagram in Figure 6.5 shows the mechanical distinction between P- and S-waves.

For earthquake sources within a few kilometers of Earth's surface, P-waves have a wave speed of about 5,400 meters per second while S-waves travel at just 3,200 meters per second. (These values are listed in Table 5.1.) This means that for each 1-second time interval, a P-wave travels an additional 2,200 meters beyond the S-wave that was generated by the same event. This is much more than a bit of scientific trivia, for it gives us a strategy for determining the distance between the point of observation and the source of the earthquake.

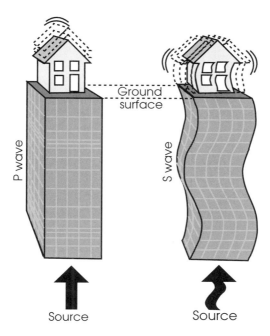

Figure 6.5. Primary (P-) and Secondary (S-) waves. P-waves are transmitted as a compressional disturbance, while S-waves are transmitted as a transverse disturbance.

Suppose, for instance, that your seismograph detects a P-wave and then, 30 seconds later, an S-wave. How far away was the source of the earthquake? Verbal logic begins to get clumsy here, because the time difference we've measured is not the same as the time the waves have actually been traveling, and the distance between the two wave fronts is not the same as the distance from our point of observation to the earthquake source. Algebraic symbolic logic provides a more convenient way to analyze this type of problem, and I invite readers with this level of mathematical fluency to write and solve the relevant equations. (My own answer is 235.6 km for a P-S time interval of 30 s.) If we are willing to sacrifice a little bit of precision, though, we can arrive at essentially the same answer through the graph in Figure 6.6. On the horizontal axis we identify the difference in arrival time of the P- and S-waves, while on the vertical axis we read the corresponding distance to the source. Accordingly, a 30-second P-S time interval corresponds to a 235-kilometer distance from the source, and a 10-second interval reflects a 79-kilometer distance.

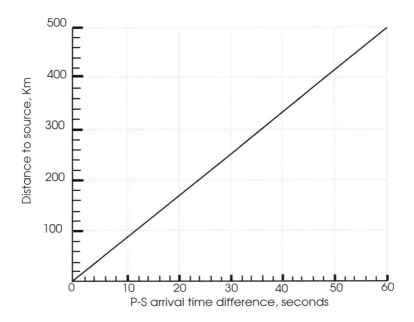

Figure 6.6. The distance to the source of an earthquake can be determined from the time difference between the arrival of the first P-wave and the first S-wave.

What this means is that we presently *do* have the ability to predict earthquakes, albeit only a few seconds in advance. If you happen to be watching a seismograph when a P-wave arrives, and the source is 79 kilometers away, you have about 10 seconds to get under a desk or into a doorframe before the potentially devastating S-wave arrives. If the source happens to be farther away, you have more time. Schoolchildren and office workers in seismically active areas are routinely drilled in how to respond to this short delay between the initial onset of the sensible tremor of a P-wave and the more destructive S-wave that follows. The basic strategy, when caught inside a building, is to quickly get under something that won't collapse on your head, and that in the worst case will surround you with breathing space until rescuers get a chance to dig you out of the rubble.

The recordings of seismographs usually appear to the untrained eye as random squiggles, and the onset of an earthquake's P- and S-waves may not be particularly apparent on an instrument that happens to be oriented in a less than optimal direction for the detection of a particular earthquake. In order to locate the source of an earthquake, scientists need to examine data recorded by multiple instruments at multiple observatories. Figure 6.7 shows a somewhat idealized seismographic recording in which the onset of the P- and S-waves both appear on the same record and are about as clear as one can hope for. Even here, it is difficult to read the P–S time interval to an accuracy much better than about half a second. This translates to an uncertainty of about 4 kilometers in distance to the source.

With modern instrumentation technology, one might expect that a greater accuracy is achievable. But it isn't. The limitation, it turns out, has nothing to do with the sophistication of the instrumentation. One cannot pinpoint an earthquake's source more accurately than within a few kilometers (or in some cases, tens of kilometers), for the simple reason that the source itself is usually this big. An earthquake does not result from an explosion of energy at a single geometrical point in the crust; it results, rather, from the sudden slippage of a section of fault line that may be many kilometers in length. This, in fact, is why Charles Richter chose to use 100 kilometers as his "standard" distance from the source in establishing his scale of magnitudes. If he'd chosen 1 kilometer or even 10 kilometers, such a reference distance would be smaller than some of the earthquake sources he was trying to measure, and his scale would have been useless as a basis for comparing different events. By choosing to use a 100–kilometer reference distance, he was assured that a source size of a few kilometers would not affect his determination of the earthquake's magnitude by more than a few percent. In fact, discrepancies of a few percent in Richter magnitude are common in the scientific literature: One researcher may describe a particular

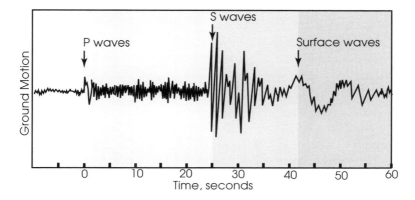

Figure 6.7. A seismographic recording of an earthquake. In this case the first S-wave arrived about 25 seconds after the first P-wave, implying that the source was about 196 kilometers from the observatory.

earthquake as a magnitude 7.7 event, while another, at a different observatory, calculates a 7.9 for the same event. This is not a matter of someone's poor measurements or miscalculations. It is, rather, a reflection of the inherent fuzziness of the phenomenon of earthquakes.

Regardless of whether one uses the Richter Scale or the moment magnitude, the "magnitude" of an earthquake always refers to the amount of energy released at the source rather than to the wave amplitudes recorded at any specific observatory. The "source," in turn, is the somewhat fuzzy region of physical origin of an earthquake; that is, the segment of the fault that has slipped. The geometrical center of this extended source region is called the earthquake's *focus;* in mathematical language it is treated as an exact point. The focus may be near Earth's surface, or it may lie well within the crust. The point on Earth's surface directly above the focus is called the earthquake's *epicenter.* The relationship between an earthquake's focus, its epicenter, and its physical source is depicted in Figure 6.8.

This geometrical idealization notwithstanding, an earthquake's epicenter is a very useful concept, because it can be pinpointed fairly accurately (typically to within a few kilometers) by combining data that have been recorded

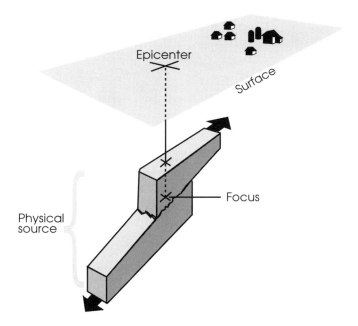

Figure 6.8. Relationship between an earthquake's epicenter, focus, and physical source.

at three different seismographic observatories. The diagram in Figure 6.9 shows how this is done. First, each observatory uses its own recorded P- and S-wave arrival times to determine its own distance from the source. This may tell Observer *A*, for instance, that the epicenter is somewhere on a 250-kilometer radius circle centered at his observatory. Observer *B*, meanwhile, might conclude that the epicenter lies somewhere on a 250-kilometer radius circle centered at that observatory, while to Observer *C* the data suggest a circle with a 350-kilometer radius. When the distances from these three observatories are combined, there remains only one possibility for the location of the epicenter. This location can be determined graphically, as in Figure 6.9, or more precisely by doing a trigonometric calculation. Although data from only three observatories are required, in practice the reliability of the computation is improved by including data from additional observatories.

The epicenter identifies the spot that is directly above the focus, or the region of fault slippage. The depth of the focus, however, is more difficult to determine. We saw in Table 5.1 that at great depths both the S- and P-waves travel at higher speeds than they do near the surface. To determine the depth, you first need to know the wave speeds, which in turn depend on the unknown depth you are trying to calculate. The usual computational

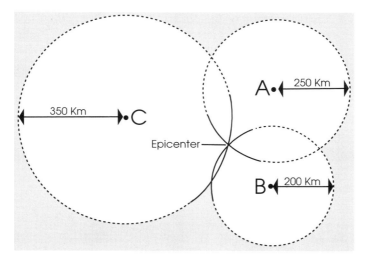

Figure 6.9. An earthquake's epicenter can be located by combining data from three seismographic observatories. In this case, the epicenter is simultaneously 250 kilometers from Observatory *A*, 200 kilometers from Observatory *B*, and 350 kilometers from Observatory *C*.

approach is one of trial and error: guessing a focal depth, computing what the various seismographic observatories ought to have recorded, comparing this with their actual recordings, adjusting the original guess accordingly, then repeating the computation with the revised value in a sequence of successive approximations. Even this process does not always generate highly consistent results. It does, however, tell us that perhaps 75% of the seismic energy released worldwide comes from foci no deeper than 10 to 15 kilometers, so that in most cases it becomes a fine distinction whether one refers to the epicenter or the focus. These shallow-focus earthquakes are, as one would expect, the most devastating. As for the other 25%, foci as deep as 680 kilometers have been identified, well within Earth's mantle. Although such deep-focus earthquakes seldom cause much surface damage, their seismological records are quite useful for mapping the structure of Earth's interior and for expanding the data bases theorists have to draw upon in their attempts to refine the current theory of plate tectonics.

Surface Waves and Structures

The major damage wreaked by earthquakes is not always due to the direct impact of the P- and S-waves that have traveled through the bulk mass of Earth. Often the destruction results from other waves that follow in complex patterns, riding upon Earth's surface. These *surface waves* (Figure 6.10), which are often dramatically visible during a strong earthquake, arise when the P- and S-waves strike Earth's surface from below.

The first surface wave to arrive, essentially on the tail of the S-wave, is called a *Love wave;* it is characterized by a lateral snaking motion of ground's surface that can be particularly destructive to building founda-

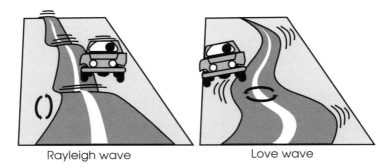

Rayleigh wave   Love wave

Figure 6.10. Surface waves.

tions, water lines, and gas lines. Shortly after the arrival of the Love wave, and traveling at about 92% of the speed of the S-wave, is the *Rayleigh wave.* This wave has a rolling up-and-down motion, quite similar to a water wave; in fact, Rayleigh waves entering an inland lake will generate water waves on the surface. Rayleigh waves often wreak havoc with bridges and elevated highways. Both Love and Rayleigh waves will set tall buildings vibrating, and will tear power and communication cables from their poles.

When surface waves encounter damp soil (particularly sand or clay), they jostle the soil particles in a way that makes the material behave temporarily like a liquid rather than a solid, a phenomenon known as *soil liquefaction.* Any structure that depends on such soil for its support will quickly sink under these conditions. The photograph in Figure 6.11 shows a row of apartment buildings that settled dramatically due to soil liquefaction during a 1964 earthquake in Japan; these buildings had to be demolished even though structurally some of them were relatively undamaged. Clearly, in regions subject to earthquakes it is always more prudent to build on rock than on sand or clay.

Figure 6.11. Soil liquefaction during the earthquake of June 16, 1964, toppled these apartment buildings in Niigata, Japan. (Photo courtesy National Geophysical Data Center.)

Building codes in seismically active areas (most of California, for instance) are designed to safeguard against the types of motion induced by surface waves. Wall framing is securely anchored to foundations with bolts that prevent base-shear failures; diagonal bracing is built into walls; garages are cross-braced to prevent torsional vibrations; and chimneys are internally reinforced with steel and tied securely to the rest of the structure. And, of course, all structural concrete must be reinforced with imbedded steel rods or wire mesh. Looking at the death tolls in earthquakes around the globe (Appendix B), there can be little doubt that casualties are highest in places that lack strict building codes. Unfortunately, it is quite difficult to retrofit an older structure to improve its resistance to surface waves, and in economically underdeveloped areas it may be very difficult to get people to comply with stringent building codes that have been adopted to protect them in an earthquake that may not actually occur within their lifetimes. Living in earthquake-vulnerable structures is not always a case of humans choosing to take chances; in many parts of the world, large segments of the population simply have no economically acceptable alternative. Such people will benefit greatly if someday it becomes possible to anticipate earthquakes and to issue timely warnings to evacuate unsafe structures.

Earthquake Prediction

We could save countless lives each year if we could answer three questions about an earthquake before it happened: Where? When? and How big?

Unless you are located right at the epicenter, an earthquake will always telegraph a few seconds' advance notice before the arrival of the most damaging surface waves. This is enough time for an astute driver to stop the car and for people caught inside buildings to crouch in doorframes or beneath sturdy tables. A few seconds, however, is not enough time to evacuate buildings, let alone cities, and in any case running into a busy street where power lines and building facades are about to topple is hardly the smartest thing to do. Clearly, what is needed is an advance notice of a few hours to a few days rather than a few seconds.

In northern California near San Francisco, a long segment of the San Andreas Fault has been locked up since the disaster of 1906; in southern California near Los Angeles, another segment of this fault has not slipped since a large (Mercalli intensity X–XI) earthquake in 1857. In between, the Pacific side of this long strike-slip fault is continually grinding northward at 2 to 5 centimeters (up to 2 in.) per year. This suggests that several sections of the fault near the most populated areas of California are capable of suddenly breaking loose and almost instantly catching up on several meters of lost movement. Such earthquakes would be catastrophic indeed, easily registering a Richter magnitude of 8 or greater. Will this actually happen? Most probably. When? Maybe a century from now, maybe tomorrow.

We have recent evidence that the stress is indeed building up, because in the last few years both San Francisco and Los Angeles have experienced damaging earthquakes due to slippages in the vast subterranean network of faults west of the boundary between the Pacific Plate and the North American Plate. At 4:31 AM on January 17, 1994, a Richter 6.6 earthquake centered 15 kilometers below the community of Northridge killed 55 in the Los Angeles area, caused several elevated freeways to collapse, plunged 3.1 million people into darkness, caused a major gas main to explode, and damaged 40,000 buildings (1,600 so badly that they had to be demolished). Just a few years earlier, on October 17, 1989, television viewers across the nation who had tuned in to watch the San Francisco Giants battle Oakland in the World Series saw their screens go blank as a magnitude 6.9 earthquake struck just south of the city. In that earthquake, 62 people lost their lives, a bayfront residential section collapsed due to soil liquefaction, and several city blocks burned out of control through the night. Yet in neither of these events was there any slippage along the locked-up portions of the San Andreas Fault. That major fault's pent-up energy is still there, and the strain continues to increase year by year.

Mathematical models have been developed that relate the recurrence interval between earthquakes to the normal slip rate of a locked-up fault and the magnitude of the earthquake that results when the fault finally does break loose.[6] For example, if a fault normally moves 1 centimeter per year but then locks up for about 100 years, it is capable of generating a magnitude 7 earthquake. On the other hand, if the fault's normal slip rate is 10 centimeters per year, and it locks up, it takes only around 50 years to build up to a devastating magnitude 8 earthquake potential. Of course, none of this actually predicts what will happen, or when; it simply sets limits on what *may* happen. Expressing this in terms of probabilities, the U.S. Geological Survey now estimates that the chance of a magnitude 8.3 earthquake along the southern section of the San Andreas Fault is somewhere between 2% and 5% per year, or about 50% over the next twenty to thirty years.

Despite their obvious limitations, such probability values are far from useless information. A series of ten magnitude 7 earthquakes will release the same total energy as a single magnitude 8 earthquake. Yet a single magnitude 8 event can cause considerably more than ten times the cumulative damage of ten magnitude 7 earthquakes, particularly if buildings in the area have been designed only to withstand a magnitude 7 event. If there is a reasonable probability of a magnitude 8 earthquake occurring, then it is prudent to establish building codes that prepare for such an event. Even long-term probabilistic predictions of earthquakes are therefore of considerable interest to both engineers and public policy planners.

But how far does one take such a strategy? There is fairly good evidence that the most intense earthquake in U.S. history may have been centered not in California or Alaska, but near, of all places, New Madrid, Missouri! Because this seismic event happened back in 1811, when that region was sparsely settled, there was little local property damage or loss of life. Yet on the Mercalli scale this event registered an intensity of X to XI near its source, and intensity V structural damage was recorded as far away as Pittsburgh, Washington, D.C., and coastal South Carolina. Even Boston, some 1,800 kilometers (1,100 mi) from the source, experienced intensity III effects.[7] This tremendous earthquake raised a tsunami on the Mississippi River, which then reversed its flow for roughly a half hour as it poured into newly formed depressions, creating several lakes that remain today. It's worth noting that the focus of this major earthquake was well within the boundaries of the North American Plate, where there is no obvious geological fault. Is southern Missouri at risk from a similar event today, more than 180 years later? Almost certainly. Not only is it at risk, but a recurrence of the 1811 event in the central United States would precipitate a much greater disaster than the "big one" everyone is currently trying to forecast in California. Most of California's buildings are designed to withstand a fair amount of ground motion; most of Missouri's buildings are not.

It's clear that earthquakes do occur in the eastern portion of North America; Table 6.4 lists the more notable ones since 1634. Eastern earthquakes seem to occur on ancient faults that can lie dormant for as long as thirty thousand years and whose surface features have long been eroded into oblivion. Over such long time spans, even probability calculations can be meaningless. It makes little sense, for instance, to design a structure to withstand a magnitude 7 earthquake when the probability of one occurring within the building's lifetime may be only a tenth of 1%. At present, we have little alternative to accepting the fact that when the next East Coast earthquake strikes, the affected region will have been caught almost totally unprepared.

Obviously, there is a great gap between our abilities (1) to establish the probability of a future earthquake within a period of decades or centuries and (2) to anticipate a surface wave a few seconds before it arrives. Considering the tremendous quantity of energy released during an earthquake, it

Table 6.4. *Some notable earthquakes in eastern North America*

Date	Place	Mercalli intensity	Comments
1638, June 11	Massachusetts	IX	Many chimneys fell.
1663, Feb. 5	St. Lawrence Region	X	Chimneys fell as far away as Mass.
1732, Sept. 16	Ontario, Can.	IX	Seven died in Montreal.
1755, Nov. 18	Massachusetts	VIII	Masonry buildings damaged. Felt from Chesapeake Bay to Nova Scotia.
1811–12	New Madrid, Mo.	XI	Three principal quakes on Dec. 16, Jan. 23, and Feb. 7. Extensive permanent changes in ground elevation and courses of rivers. Damage as far as Cincinnati and Richmond. Felt in Boston.
1870, Oct. 20	Montreal and Quebec	IX	Damage also reported in Maine.
1886, Aug. 31	Charleston, S.C.	X	60 killed, 102 buildings destroyed, 90% of buildings damaged. Felt in Boston, Chicago, and St. Louis.
1895, Oct. 31	Missouri	IX	Felt from Canada to Louisiana.
1909, May 26	Aurora, Ill.	VIII	Numerous chimneys fell.
1929, Aug. 12	Attica, N.Y.	IX	250 chimneys fell.
1929, Nov. 18	Off Grand Banks, Newfoundland	X	Broke 12 transatlantic cables, some as far as 250 km apart. Several deaths from tsunami.
1931, Apr. 20	Lake George, N.Y.	VIII	Chimneys fell.
1944, Sept. 5	Canada and New York State	IX	Destroyed or damaged 90% of chimneys in town of Massena

Note: For a more extensive list of earthquakes, see Appendix B.

seems unlikely that there would be no precursors in an intermediate time interval – say a few hours or days in advance. Break a stick over your knee, and you'll hear it begin to fracture just before it snaps. Twist the cap off a fresh jar of pickles, and you'll feel the cap begin to creep just before it releases. Surely, it would seem, there must be similar precursors in the hours, days, or months before a locked-up geological fault breaks loose catastrophically. If so, the scientific discovery of such precursors (and by "scientific," I mean quantitative and replicable) could someday lead us to develop a practical method for forecasting earthquakes.

The hunt for reliable precursors has been going on for several decades, and scientific journals have published many articles suggesting that earthquakes may indeed whisper subtle messages before they unleash devastating amounts of energy. Water levels in wells sometimes fluctuate just before an earthquake, and occasionally radon gas is released. There is enticing evidence that some faults may emit a burst of long radio waves shortly before they rupture. There is also the phenomenon of diletancy: the slight temporary expansion of a stressed material just before it settles into a new and more stable configuration. Although a direct measurement of diletancy requires a very sensitive and properly placed tiltmeter, an indirect measurement is sometimes possible by measuring the wave speeds of S-waves that originate from other sources and travel through the region of interest. Rock that is stressed near the point of rupture will split an S-wave into two components that travel at slightly different wave speeds and therefore have different times of arrival at a seismograph. The data analysis, unfortunately, is quite difficult and fraught with uncertainties, and this technique so far has succeeded mainly in predicting earthquakes that did not materialize.

Some scientists have also examined animal behavior. Throughout history, there have been many anecdotal reports of cattle being spooked, coyotes howling, dogs barking wildly for no obvious reason, ducks scampering out of ponds, hogs becoming uncharacteristically quiet, and chickens flying into trees shortly before an earthquake. In Japan, one sixteen-year study (abandoned in 1992) was explored the use of catfish to predict earthquakes. For twenty-four hours a day, the movements of these sluggish and weak-eyed fish were monitored electronically and compared with measurements of seismic activity. Although the fish indeed perked up a few days before about 31% of the earthquakes, their behavior failed to correlate at all with the earthquake magnitude, and the prospects seemed slim for ever basing a practical earthquake warning system on such an unreliable biological phenomenon.

The main problem in doing this type of research is that we don't know where best to locate the instrumentation (or the catfish) until *after* an earthquake has struck, which of course is much too late to detect any precursors. The vast majority of earthquakes occur at sites where no one has ever had any reason to install instrumentation in the immediate area. Once again, a Catch-22: To predict earthquakes, we need to learn what precursors to look for, but to get instrumentation in place to identify the precursors, we first need to predict where and when an earthquake will occur. It simply isn't practical to cover the whole state of California with tiltmeters, radon detectors, and/or Japanese catfish, then watch them all. Clearly, scientific progress in this area is going to need a helping hand from Lady Luck.

Yet it's still much too soon to become pessimistic about the prospects of someday identifying reliable precursors. Although earthquakes have killed and maimed throughout human history, only relatively recently (within the twentieth century) have we made any notable progress in measuring and classifying seismic events. The science of seismology is still very young, and as with any young science its first accomplishment has been to identify the classes of questions that beg for further scientific study. From an academic perspective, then, this is far from a dead-end field for future scientific inquiry.

Modern science, however, is driven not merely by questions of academic interest, but by the questions our society most desperately needs to have answered. Clearly, with populations expanding rapidly in seismically active regions of the globe, each year an increasing number of human lives becomes hostage to our inability to forecast earthquakes a few hours or days in advance. The social price of our scientific ignorance escalates year by year, while the cost of funding additional scientific research plummets relative to its potential social benefits. If future governmental investment in this line of research should in fact result in a reliable method of forecasting earthquakes, it would turn out to be at least as big a bargain as was the investment in the development of the polio vaccines.

Notes

[1] D. Landen, Alaska earthquake, *Science* (1964)

[2] B. A. Bolt, *Earthquakes* (New York: Freeman, 1988).

[3] B. F. Howell, Jr., *An introduction to seismological research* (Cambridge: Cambridge University Press, 1990).

[4] C. Davison, *The founders of seismology* (Cambridge: Cambridge University Press, 1927).

5   An earthquake in this region in December of 1920 claimed some 100,000
    lives. For an account, see U. Close & E. McCormick, Where the mountains
    walked, *National Geographic,* May 1922, 445–64.

6   L. Reiter, *Earthquake hazard analysis* (New York: Columbia University Press,
    1990).

7   O. W. Nuttli, The Mississippi Valley earthquakes of 1811 and 1812: Intensi-
    ties, ground motion and magnitudes, *Bulletin of the Seismological Society of
    America,* 63 (1973), 227–48.

7

Volcanoes and Asteroid Impacts

St. Pierre, Martinique, 1902

This city of thirty thousand exists no more. Devastated cities are rebuilt by their survivors, and at St. Pierre there were but three. Two died soon after being rescued. The third, a convicted murderer liberated from an underground jail cell three days after the disaster, recovered from his burns and emigrated to the United States to live out his years as an attraction in the Barnum & Bailey Circus. Today, at the site of the 1902 catastrophe, one finds a modest museum, trees growing in roofless ruins, and a small nearby settlement that has grown in recent decades but bears no resemblance to the vibrant population center that flourished here at the turn of the century.

In 1902, St. Pierre, Martinique, was the gem of the French West Indies, its picturesque two- and three-story houses roofed in red tile and painted in bright tropical colors, its gardens and courts landscaped with lush tropical plants. The city was far from level; steps had been cut into many of its steeper streets, and portions of the harbor dropped so steeply into the sea that large ships could anchor within a stone's throw of land (Fig. 7.1). Although Martinique's political capital lay 18 kilometers (11 mi) to the south at Fort-de-France, St. Pierre was effectively the social and commercial capital, its thriving economy sustained by the numerous sugar plantations that dotted the 32 km by 72 km island. Napoleon's empress Josephine had been born here, and in France the city was often referred to as the "little Paris."

Some 7 kilometers to the north of the city towered Mont Pelée, its summit cradling a lake 1,350 meters (4,428 ft) above the sea that provided a popular site for swimming and picnicking. Twenty-five streams cascaded down the slopes of this ancient volcano, several in the general direction of St. Pierre. One river emptied into the sea a bit north of the city, a second ran through a ravine that bounded the city on the north, and another shallower and more turbulent stream gurgled through the center of the city, its high banks lined with grand villas. The basic geography of the region is shown in Figure 7.2.

At 8:02 AM, on the morning of May 8, 1902 (the time is based on an interrupted message at the cable office at Fort-de-France), Mont Pelée exploded catastrophically and unleashed a "whirlwind of fire" toward St. Pierre. Within two minutes of the explosion, this avalanche of incandescent ash and gases engulfed the city, flattened most of the buildings, and set fire to the debris. In the harbor, the blast destroyed eighteen sizable ships and an unknown number of smaller boats.[1] Only the British steamship *Roddam* survived, after sustaining serious fire damage and the loss of two-thirds of its crew.

Compared to earthquakes, volcanoes are considerably more generous about giving advance warning of an impending disaster. Unfortunately, reading these warnings correctly is akin to predicting when a sleeping tiger will awake by interpreting its snores: An error in judgment may not be retractable. Mont Pelée slept soundly between 1851 and 1902.[2] On that earlier date the volcano expelled a cloud of smoke and steam that rained a few centimeters of ashfall onto St. Pierre, but there were no deaths and no structural damage, and the lush tropical vegetation quickly recovered. The vol-

Figure 7.1. St. Pierre, Martinique, in a photograph taken a year or two before the 1902 disaster.

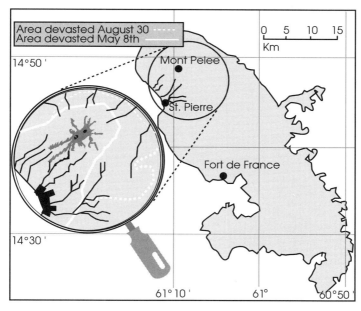

Figure 7.2. The geography of Martinique and the region surrounding St. Pierre, showing the areas devastated by the *nuees ardentes* of May 8 and August 30, 1902.

cano didn't stir again until February of 1902, when residents first noticed faint rumblings and emissions of steam, and everyone prudently quit taking their recreational hikes to the summit. The thunder from the mountain continued irregularly over the next few months, and from time to time huge ash clouds billowed into the sky. In late April it became fairly common for these mushrooming clouds to expand to proportions that darkened the midday sun. By now, the air carried the unmistakable smell of sulfur oxides, dead birds were being found in some of the ashfalls, and city residents began wearing wet handkerchiefs over their faces. On April 27, a small expedition climbed to the summit and returned to report that a previously dry ravine had been transformed into a lake and that the rumbling volcano had grown a new cinder cone. As horses began dropping dead of asphyxiation in the streets of St. Pierre, a series of small earthquakes broke undersea telegraph cables. On April 30, without warning, the benign little streams leading into town suddenly became raging torrents, carrying boulders and tree trunks from the upper slopes of the mountain and claiming several victims. On May 2, the ashfall in portions of the city had accumulated to a depth of 40 centimeters (16 in.), and the local newspaper announced that a new expedition would climb to the summit on May 4. Then, just before midnight on May 2, the city was awakened by a violent series of ground tremors and brilliant flashes of lightning in the towering ash clouds. By Sunday morning, May 4, the entire harbor was littered with dead birds, and none of the members of the announced expedition was foolhardy enough to go through with the plans to climb the volcano for a closer look at what was happening.

The volcano calmed down on the morning of May 5, but shortly after noon (before anyone could breathe much of a sigh of relief) the sea in the harbor withdrew 100 meters, then swelled to flood the waterfront area of the city. This water wave was not, it turns out, a conventional tsunami; rather, it seems to have resulted from the collapse of one of the walls that held the crater lake at the summit. A surge of boiling water and mud, trailing a great plume of steam, had thundered down one of the gorges, demolished the largest sugar factory on the island, and buried 150 victims in up to 100 meters of mud before the avalanche hit the harbor. Most likely, the withdrawal of the harbor waters was preceded by a swell that was not reported in the surviving accounts.

At this point the city was in pandemonium, with refugees funneling in from the suburbs and colliding in the narrow streets with residents attempting to evacuate. The governor called upon a committee of experts (whose credentials do not seem to have been recorded) for a recommendation as to whether the city should be evacuated. These experts reported, "The relative positions of the craters and valleys opening toward the sea sanction the conclusion that St. Pierre's safety is not endangered." Apparently the committee's reasoning was that any major lava flow would be stopped, deflected, and/or channeled by the ravines north of the city, and that the now-continuous ashfall was merely a nuisance. The governor had more than a casual interest in keeping the residents of St. Pierre in place, for an election was coming up on May 10, and people who are displaced from their homes do not put voting high on their list of personal priorities. To instill public confidence in the expert report, Governor Mouttet and his wife moved from Fort-de-France to St. Pierre on May 6. It was a bad decision; two days later they would both be incinerated, along with the rest of the 30,000 inhabitants. Volcanoes are not inclined to show much respect for human political concerns.

At 4 AM on May 7 – St. Pierre's last day – Mont Pelée began roaring. Lightning flashed continually around the summit, and two volcanic vents blew giant fountains of glowing cinders into the predawn sky. The surrounding sea was black with ash. More people fled the city, while additional refugees crowded in, the numbers tipping in favor of the refugees and actually swelling the ultimate death toll. On the last day its presses would run, the local newspaper editorialized, "Where better off could one be than in St. Pierre?" Unimpressed, the captain of an Italian ship steamed out to sea after loading only half his cargo. His home port: Naples, in the shadow of Mount Vesuvius – the famous volcano that had buried Pompeii in A.D. 79. This captain was not about to trust his life to the local experts; he knew quite well what destruction a thundering volcano is capable of unleashing.

The cataclysmic explosion on May 8 was witnessed from Fort-de-France, from a few mountain villages that lay outside the direct path of the blast, and from several ships that were at sea beyond the harbor. Although the accounts differ in minor details (particularly the timing), they substantially agree in their general descriptions of what happened that terrible morning.

There were two blasts, virtually simultaneous. One exploded straight up, sending a great billowing cloud of ash 11,000 meters (7 mi) into the atmosphere. The second was a *nuees ardentes,* or pyroclastic flow, that burst down Mont Pelée's southwestern slope at a speed of around 190 kilometers per hour (120 mi/h). In two minutes or less,[3] this ground-hugging cloud of superheated gas and ash engulfed everything that lay between the volcano and the city, its frontal surface a billowing mass so hot in places that it glowed incandescently. The ravines north of the city did nothing to deflect this blast, and it was quite impossible for anyone to outrun it. On impact, it crumbled meter-thick masonry walls in St. Pierre and capsized large steamships in the harbor. The high temperature (estimated at 700–1,200 °C) instantly cremated most of the victims while igniting the splinters of the city and the wooden decks of the ships that were still afloat. No sooner had the pyroclastic flow passed than a gale-force wind arose in the opposite direction, as if there were a great vacuum to fill. This wind displaced the volcanic gases with fresh air and further fanned the flames. Huge flashes of lightning danced in the ash clouds above the demolished city, and for the next hour it rained, covering everything with a thick paste of volcanic mud. Even this rain, however, could not extinguish the widespread fires; they were still smoldering four days later.

A ship full of officials and military personnel was dispatched from Fort-de-France, and it arrived at St. Pierre around 12:30 PM, less than five hours after the disaster. The harbor was now a small sea of floating volcanic ash, littered with the debris from burning and capsized ships and hundreds of charred corpses. Although a number of seamen and ship passengers were pulled from the water alive, few of these victims ultimately survived their burns. Onshore, the ground surface was so hot it repelled attempts to land for several hours. At this point, however, there was no longer a great deal of urgency to get ashore, for the view from deck already made it clear that the city of St. Pierre would be in no need of emergency aid. Thirty thousand had perished, and there was nothing left to save or even salvage. The investigators did, however, eventually tramp through the ruins to search for survivors, and found the three I mentioned earlier. Figure 7.3 is a contemporary photograph of the devastation, with Mont Pelée in the background.

Although St. Pierre's human disaster was complete, the geophysical event was not. Mont Pelée blew its top again two weeks later (May 20) and toppled

From Orange Hill (N. E.) over dead St. Pierre, to steaming Mont Pelée—once beautiful and verdure-clad—Martinique.   Copyright 1902 by Underwood & Underwood.

Figure 7.3. The destruction at St. Pierre, in a photograph taken shortly after the 1902 disaster. (Photo courtesy Library of Congress.)

what few walls remained standing in the city. Then, on August 30, 1902, the volcano dispatched another pyroclastic flow slightly to the east and incinerated the village of Morne Rouge and several adjacent hamlets, claiming 1,500 to 2,000 additional lives. These later explosions seem to have equaled or exceeded the violence of the May 8 event.

In October, observers reported a spine of lava with a diameter of 150 meters (500 ft) growing from the floor of Pelée's crater at a rate of up to 10 meters per day, an amazing speed for a geological phenomenon. This "Tower of Pelée" reached a height of 311 meters (1,020 ft) before it collapsed and started to grow again, and in one day (August 31, 1903) it grew a reported 24 meters (78 ft)! After 1904, the tower disintegrated until it amounted to but a stump in the middle of a heap of rubble. Such lava dome formation is usually an indication that a volcano is in its final stage of eruption, when all gases that can cause a major explosion have already been expelled from the subterranean chamber of hot magma.

Yet on September 16, 1929, Mont Pelée began to rumble once more, and again it discharged several pyroclastic flows. This time there was no loss of life, because all of the 1,000 nearby residents had wisely heeded the precursers and evacuated. By late 1932, the volcano settled down again. It has been quiet ever since.

St. Vincent Island, 1902

One peculiar aspect of Mont Pelée's deadly eruption of May 8, 1902, is that a nearly identical volcanic disaster had claimed 1,350 lives only 160 kilometers (100 mi) distant *just one day earlier.* At 2 PM on May 7, 1902, the volcanic mountain La Soufrière exploded and devastated 115 square kilometers (45 mi^2) of the northern end of the island of St. Vincent. The mechanism of destruction was the same: a ground-hugging cloud of superheated gases expanding at breakneck speed: a *nuees ardentes,* or pyroclastic flow.

Between the 1600s and 1902, there were only two major volcanic eruptions in all of the Lesser Antilles, and both were at La Soufrière on St. Vincent. In 1718, this volcano smothered the entire island and much of the surrounding sea under a huge ashfall. In the 1812 eruption, La Soufrière blew so much ash into the atmosphere that it plunged the island into a full day of total darkness. As the crater from the 1812 eruption filled up with rainwater, it grew into a lake nearly 1 kilometer in diameter and 175 meters deep, its surface 1,100 m (3,500 ft) above sea level – amazingly similar to Mont Pelée's crater lake.

La Soufrière began stirring again in April of 1901, about ten months before the first hints of Mont Pelée's renewed activity. By the beginning of May, most residents in the shadow of La Soufrière were wisely evacuating to the southern end of St. Vincent. By 10:30 AM on May 7, the noise of the eruption had become an almost continual roar, and a huge steam cloud billowed upward to an altitude of more than 9,000 meters (30,000 ft). By 1 PM, airborne boulders could be seen in the cloud, and a normally dry riverbed suddenly became a raging torrent of boiling mud and water more than 15 meters (50 ft) deep. This *lahar* put an end to all prospects for additional southward evacuations on the eastern side of the island.

An hour later, La Soufrière cut loose with the huge *nuees ardentes* that incinerated 1,350 human victims. Unlike Mont Pelée, La Soufrière's devastating blast swept outward in all directions from the volcano (a fact confirmed later by the radial pattern of the fallen trees); as a result, its energy seems to have dissipated more rapidly. In fact, the eastern portion of the pyroclastic flow may even have stopped and reversed direction back toward the volcano, driven by an inrush of air flowing into the partial vacuum left behind by the explosion. In spite of this quick dispersion, the *nuees ardentes* left huge accumulations of ash, portions of which remained hot for weeks. The moderate death toll of 1,350 (compared to St. Pierre's 30,000) reflects the geographical accident that no large city lay in the direct path of the most intense portions of the pyroclastic flow. The volcano continued to erupt intermittently through the month of May, and again from September 1 through 3, 1902. It then remained relatively quiet until the 1970s.

What role did the St. Vincent disaster play in warning the people of St. Pierre, less than 160 kilometers to the north? Apparently none whatsoever. The languages were different (French in Martinique, English in St. Vincent), and in any case most of the undersea telegraph cables in both areas had been severed by earthquakes. Can we connect the two events geophysically? Yes, in terms of the plate tectonics of the region; no, in terms of the fact that they occurred just a day apart. In geophysical time, there is little difference between a day and a century. We can, however, make the reasonable speculation that both Mont Pelée and La Soufrière released energy from the same tectonic source, and that had there been but one active volcano in the region at the time rather than two, the single volcanic explosion would have been considerably greater in magnitude than either of the two historical events.

There is relatively little literature on the 1902 St. Vincent event, apparently because the scientists of the time were drawn to the greater human impact of the destruction of St. Pierre that occurred the next day. With the benefit of hindsight, we can view this as a missed opportunity for contemporary scholars to compare and contrast the documented precursors and consequences of these two very similar events.[4] Never before or since have two separate volcanoes caused such human devastation in such close proximity in space and time.

The Mechanisms of Volcanism

Major volcanic disasters are relatively rare in the course of human events; in fact, quite a few large cities have stood unharmed for many centuries in the shadow of an active volcano. Worldwide, there are only a handful of major eruptions in any typical year, and these usually occur in sparsely populated regions. Volcanoes around the globe claim human victims only a few times in a decade, and they cause major devastation only a few times per century. But when they do, there is considerable terror in the event – for there is nothing on earth more indiscriminate in its destruction than a volcano on a rampage.

Scientists usually consider a volcano to be active if it has erupted in the last ten thousand years, roughly the period of time since the last ice age; according to this criterion, there are some 1,343 active volcanoes above the sea[5] and an untallied but certainly comparable number below the oceans. Today, at least 500 million people dwell close enough to such volcanoes for their lives to be threatened by an eruption. Meanwhile, there is evidence that some volcanoes may stay intermittently active for up to 10 million years. It is impossible to place a sure bet that any volcano is totally extinct.

The most violent volcanoes are found in the world's most earthquake-prone regions: around the rim of the Pacific Ocean, and along an arc that extends from the Mediterranean Sea to Iran and (after a gap) continues through Indonesia to the western Pacific. The theory of plate tectonics helps to clarify this observed link between earthquakes and volcanoes. Near the boundary where one crustal plate slides beneath another, the lifted plate is free to relieve its stress through earthquakes while the subducting plate is subjected to an ever-increasing compressive stress and (at a depth of around 100 km) begins to liquefy. Any weakness in the rising plate provides a path for the molten rock of the subducting plate to relieve its pressure by flowing upward to the surface. The result is a *subduction volcano,* which generally occurs roughly 200 kilometers inland from the boundary of the colliding tectonic plates. Such volcanoes are characterized by tall cinder cones and long periods of dormancy punctuated by violent explosions. Mont Pelée, La Soufrière, and Krakatau (soon to be discussed) fall into this category.

A different type of volcano arises at the boundaries where Earth's crustal plates move in opposite directions. Such regions, where the crust is slowly being ripped apart, are called *rifts;* as a rift opens, magma from Earth's interior rises to fill the opening in a relatively gentle flow. *Rift volcanoes* are found mainly in Iceland, and along a north-south line under the mid-Atlantic, and in the rift valley of eastern Africa. Rift volcanoes expel large quantities of molten lava but little airborne ash and no pyroclastic flows. They seldom kill, but they do claim property.

There is a third type: a *hot-spot volcano.* The classic example is seen in the geography of the Hawaiian island chain. If we look at a map of Hawaii, we find a string of islands running from the southeast to the northwest, which is the same as the direction of movement of the Pacific Plate. The northwestern islands are the oldest geologically, and their volcanic mountains have not erupted in millions of years. Meanwhile, the southeastern most island of Hawaii has two active volcanoes, one of which (Kilauea) is the most continuously active volcano on earth. Offshore, another volcano to the east is busily working beneath the sea to build a new island in the Hawaiian chain; it will probably emerge from the ocean surface within a few centuries. The geographical pattern here, in the middle of the Pacific, suggests that there is indeed a "hot spot" beneath the Pacific Plate that pokes its way through the undersea crust as the plate slides continually to the northwest.

These three types of volcano (subduction, rift, and hot-spot) are shown in the diagrams and photographs in Figure 7.4. Of the three, the subduction type is by far the most dangerous, its explosive potential resulting from the fact that the ejected material comes from so deep within Earth. Under a great enough pressure, molten rock is capable of holding great quantities of dissolved gases like carbon dioxide, steam, and sulfur dioxide. When this magma rises through a volcanic vent, the pressure is relieved, and the dissolved gases bubble out almost instantly. The explosion of a subduction volcano has a lot in common with popping the top on a can of warm beer after shaking it, except that the volcano is many orders of magnitude larger and hotter.

Volcanologists now know a fair amount about the underground processes that drive volcanic eruptions, and, like the seismologists, they are valiantly struggling to develop theories that will enable them to forewarn vulnerable population centers. So far, their track record is mixed. When La Soufrière began rumbling most recently in 1975, volcanologists predicted a repeat performance of the 1902 disaster, and officials evacuated 72,000 people for a period of three months. Rather than exploding, however, the volcano died down. On the other hand, similar warnings to Colombian officials in 1985

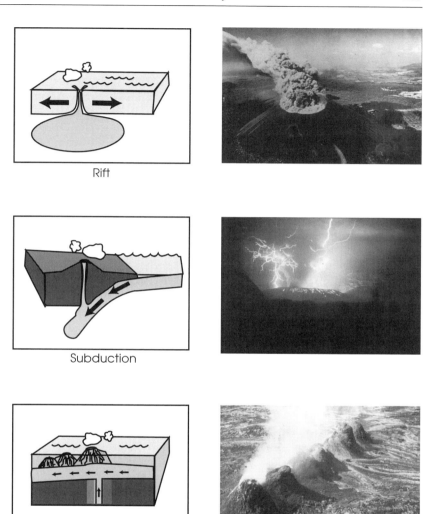

Figure 7.4. Subduction, rift, and hot-spot volcanoes. (Photos courtesy National Geophysical Data Center.)

were apparently not strong enough; the Nevado del Ruiz volcano erupted, just as many scientists were predicting it would, and 22,940 died in the lahars. One success was at the Colo volcano in Indonesia, where, within weeks after all 7,000 inhabitants were evacuated in 1983, the volcano did in fact blow its top, devastating much of the tiny island with its pyroclastic flows but claiming no human victims.

The Effects of Eruptions

Volcanoes can wreak havoc by drawing from a whole bag of tricks, summarized in Table 7.1. Although the first phenomenon listed, the ashfall, may not sound particularly threatening, near the source of an eruption the airborne ash is usually hot and wet, sticking to everything until it solidifies; it can also be acidic enough to cause acid burns on exposed human flesh. Falling ash may also be accompanied by a high concentration of carbon dioxide that invisibly fans out along the ground and asphyxiates victims before the ash itself buries them. During excavations at Pompeii in Italy in the nineteenth century, archaeologists detected hollow cavities within the solidified ash and filled them with plaster before digging farther; in doing this, they created casts of dozens of human victims, preserved as they were at their moment of death nearly two thousand years earlier. Many had their hands or clothing held to their mouths, obviously trying to keep out the suffocating dust and gases. The discovery of uneaten food on household tables suggests that more than a few of Pompeii's residents in A.D. 79 continued life as usual right up to the very end, erroneously assuming that a little bit of ashfall and gas from Mount Vesuvius were nothing to worry about.

When the word "volcano" is mentioned, most people's mental imagery turns to lava flows, those dramatic red-hot rivers of melted rock. Lava flows, however, seldom move fast enough to catch human victims by surprise; once they leave the mouth of a volcano, they follow the terrain, and their downhill course is fairly predictable. They are nevertheless terribly destructive of human structures and farmlands that lie in their path. As a lava flow approaches, it sets ablaze everything combustible; it then engulfs the remainder and solidifies everything within a mass of jagged black rock. In a few cases, slow-moving lava flows have been successfully diverted by spraying the leading edge with water to solidify the lava into a dam, while bulldozing an alternate path for the rest of the flow to follow. This strategy, however, requires a significant engineering effort (and, obviously, advance preparation), and the effort is not without its human risks. In most cases, it

Table 7.1. *Volcanic phenomena that generate human disasters*

Volcanic phenomenon	Archetypal disaster
Ashfall	A.D. 79. Eruption of Mt. Vesuvius buried the Roman city of Pompeii in up to 10 meters of ashfall and claimed up to 10,000 lives.
Lava flow	1943. A new volcano sprouted from a cornfield in southwestern Mexico, and within two years lava flows destroyed the towns of Paricutin and San Juan de Parangaricutiro.
Pyroclastic flow (*nuees ardentes*)	1902. Gaseous explosion from Mont Pelée, Martinique, incinerated the city of St. Pierre and killed 30,000.
Mud slides/floods (lahars)	1985. Nevado del Ruiz, Colombia. Heat from the volcano generated an avalanche of mud and glacial meltwater that killed 22,940 in towns as far as 50 km from the source.
Tsunamis	1883. Explosion of Krakatau in the Straits of Sunda west of Java generated tsunamis that killed 36,000.
Global weather	1816. The "year without a summer" in New England and western Europe resulted from the previous year's explosion of Tambora in Sumbawa, Indonesia. The immediate event killed 12,000; the resulting famines claimed at least 90,000 lives around the globe.

is quite useless and dangerous to attempt to hold one's ground against an approaching lava flow.

A pyroclastic flow, or *nuees ardentes*, is a billowing mass of superheated gas and suspended ash particles that expands explosively at a tremendous speed (up to several hundred kilometers per hour). Because its average density is greater than that of the surrounding air, the lower boundary of a pyroclastic flow races along the ground, knocking down and incinerating everything in its path. Pyroclastic flows have even been known to travel significant distances over water, transmitting the devastating effects of a volcanic explosion from one island to another. Although we know that this phenomenon is associated only with subduction volcanoes, not every such eruption generates a pyroclastic flow. Even so, the possibility of a pyroclastic flow needs to be considered very seriously any time a subduction volcano begins to erupt.

Mudslides and floods (*lahars*) are other possible side effects of volcanic eruptions. If the volcano has a snowcover that melts from the heat, torrents of ash-saturated water may cascade down its slopes. If there are lakes at the higher elevations, the earthquakes associated with an eruption can fracture the natural dams that impound the water. Great devastation can result at lower elevations from the sudden release of an upstream natural reservoir held in the crater of a volcano.

If a volcanic explosion occurs at sea level, a significant portion of the energy released is always transmitted to the sea water in the form of a tsunami. We have already considered the case of the Thera eruption of 1626 B.C., and we will shortly examine another case: that of Krakatau in A.D. 1883. By setting off a tsunami, a volcano can do great damage hundreds of kilometers from the source of the eruption.

The effects of a volcano, however, can reach even farther than its tsunamis. Whenever a volcano is energetic enough to blow ash into the stratosphere, Earth's jet streams whisk the particulates around the globe, and for the next few months to a year, everyone in the affected hemisphere (Northern or Southern) witnesses dramatic and colorful sunsets. Given a sufficient injection of this airborne ash, the upper atmosphere begins reflecting an unusually large amount of sunlight back into space and upsets the planet's energy balance. Then the average global temperature drops, patterns of climate are altered in complicated ways, and some regions experience crop failures and famine while other regions get increased rainfall or more temperate weather. The specific effects, region by region, are quite impossible to predict from our currently incomplete scientific models of climate. What we do know is this: A major volcanic eruption has the capacity to affect a significant fraction of Earth's population. We know this because it actually happened, in the years immediately following 1815.

Tambora, 1815

Mt. Tambora dominates the remote northern end of the island of Sumbawa, one of the great string of Indonesian islands that stretches from Sumatra on the west to New Guinea on the east. This chain of tropical islands (Fig. 7.5) is subject to intense volcanic activity along its entire 5,000-kilometer (3,000-mi) length, a result of the subduction of the northeastward-moving Indian

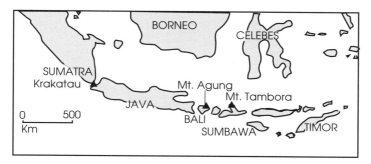

Figure 7.5. Volcanically active Indonesia. The region shown here has at least fifty active volcanoes.

Plate beneath the southeastward-moving China Plate. When Mt. Tambora yielded to these relentless tectonic stresses on April 13, 1815, it exploded with the most violent volcanic eruption in historical times, and possibly in the last ten thousand years.

Even today, the region around Mt. Tambora is still an obscure portion of the world. In 1815, long before the invention of the telegraph, the local inhabitants were almost completely isolated from the world's major population centers. No official census had ever been made prior to the disaster, and the often-quoted death toll of 12,000 is quite unofficial. (Some writers claim that as many as 90,000 may have died in the immediate event.) Although 26 natives of Sumbawa apparently did survive the disaster,[6] none of their accounts was ever written down. There is nevertheless considerable evidence from other sources that this was an eruption of stupendous proportions.

We do know that the great volcano began to rumble on April 5, and that for the next week the explosions were heard at British and Dutch colonial settlements as far as 1,560 kilometers to the west and 1,150 kilometers to the east. The midday sky was blackened within a radius of several hundred kilometers, and great mats of floating ash and pumice soon blocked the seaways. Several small tsunamis flooded harbors and beached anchored ships on nearby islands. Then, on April 13, in a cataclysmic explosion, Mt. Tambora ejected 150 to 180 cubic kilometers (36–43 mi^3) of pulverized rock and ash, reducing the height of the mountain by some 1,280 meters (4,200 ft). The volume of material launched into the atmosphere was at least 10 times as great as in the Thera eruption of 1626 B.C. Because the explosion took place well above sea level, this giant blast did not generate a major tsunami; almost all of the expelled material went upward, and most of what went up did not come down again for around a year.

It was thirty-two years later, in 1847, that an expedition finally climbed the mountain and drew the first detailed maps that permitted estimates of the magnitude of the explosion. Not until 1913 did anyone repeat the ascent, and by 1947 other explorers reported that a lake had filled the crater. Meanwhile, several violent eruptions of other volcanoes in the region allowed geologists to develop a more generalized picture of the behavior of Indonesian volcanoes. We now understand that this string of subduction volcanoes is particularly prone to blasting great quantities of ash into the upper atmosphere, where the jet streams of the stratosphere quickly distribute it around the globe.[7]

The year after Tambora's great eruption, 1816, came to be known in New England as the "Year without a Summer." Temperatures recorded at Yale University in Connecticut averaged at least 7 °F less than normal; it snowed in June in western Massachusetts and points north; and a series of frosts in June, July, and August caused crop failures throughout the northeastern United States. The summer was also uncharacteristically dry, which compounded the farmers' difficulties. (Although a 6-inch summer snowfall may be dramatic, it equates to only roughly a half-inch of rain.) The following winter was one of extreme hardship, and thousands died in the famine.[8]

Much of Europe experienced a similar cold summer and its consequent crop failures. In Germany, efforts were made to prohibit the distillation of alcohol to save the meager grain supplies for food. Grain prices nevertheless tripled, and food riots were reported in France, the Netherlands, and Switzerland. Cat and dog populations plunged, people began eating things they'd never before eaten, and in Switzerland authorities issued instructions on avoiding poisonous plants (which aren't even very plentiful in that Alpine country). Ireland suffered a terrible famine, which rendered the population particularly vulnerable to an opportunistic outbreak of typhus. Between 1817 and 1819 more than 1.5 million Irish were struck by the disease, and 65,000 of these victims died.

Tambora's eruption of 1815 never did make the newspapers in the United States or Europe, and only within the last few decades have scientists linked this great volcanic explosion to the subsequent period of bitter weather on the opposite side of the globe. There is no way today to accurately assess the full human impact of an event of 180 years ago that was not even well documented at the time it happened. At least 90,000 people probably died in the famines – considerably more if one counts victims of the collateral diseases and epidemics. Regardless of the uncertainties in quantifying such effects, it's clear that this explosion was responsible for a great deal of human misery and death around the globe. The lesson of Tambora is that a major volcanic eruption cannot be too far away geographically to be a potential threat to human life and livelihood.

Krakatau, 1883

The small, uninhabited volcanic island of Krakatau stands in the Straits of Sunda between Java and Sumatra, about 1,400 kilometers (850 mi) west of Mt. Tambora. British maps from the nineteenth century identify the place as "Krakatoa," and this spelling has found its way into much of the popular literature. In 1883, the island consisted of three volcanic cones that had grown over many centuries from a prehistoric caldera, remnants of which survived as two crescent-shaped outer islands. After August 27, 1883, only half of one of the three volcanic cones remained, neatly bisected down its middle. More than 20 cubic kilometers (5 mi^3) of Krakatau had been blown into the atmosphere, leaving an undersea crater up to 290 meters (950 ft) deep in place of two-thirds of the original island.[9] The resulting pyroclastic flows and tsunamis claimed at least 36,000 lives on nearby shores.

We know a great deal more about this event than one might guess, given its date and unfamiliar-sounding location. The Straits of Sunda were a well-traveled shipping lane by 1883, and no ship passing between the South China Sea and the Indian Ocean could avoid taking at least an informal bearing on Krakatau. There were a number of thriving Dutch colonial settlements on the nearby shores of Java and Sumatra, which led to a continuity of record keeping. Especially significant, however, is the fact that most of the world's major cities had just recently been linked through undersea telegraph cables. This made it possible, for the first time in history, to correlate observations that were made nearly simultaneously at different places on the globe.

The event began with a series of minor earthquakes on May 10, 16, and 18. On the twentieth of May, Krakatau's smallest and youngest volcano erupted with a series of explosions that sent an ash cloud more than 10 kilometers into the air. From this point on, the eruptions were nearly continuous, and by the end of the month giant volcanic clouds billowed to altitudes of around 20 kilometers (12 mi) (Fig. 7.6). Ships traveling through the straits were quickly blanketed in a thick paste of fine white ash, and sailors shoveled their decks in a gloom punctuated by flashes of lightning that danced in the volcanic clouds. On August 1, floating pumice was reported as far as 1,900 kilometers (1,200 mi) to the west. A week or so later, Krakatau's second volcano joined in with its own eruption. On August 11, a government

Figure 7.6. Krakatau, as it appeared during the early stages of its 1883 eruption, from a contemporary drawing. (Photo courtesy P. Hedervari, National Geophysical Data Center.)

surveyor landing briefly on the upwind side of the island noted that all vegetation there had been destroyed, and that there were now three separate eruption columns and at least eleven active steam vents. He wisely chose to leave before completing his survey. By August 21, unusual sunsets were reported in South Africa.

Up to this point, most of the region's colonial inhabitants viewed the event as a dramatic display of natural pyrotechnics but not as a significant threat. The nearest settlement, Kalimbang, was 30 kilometers (19 mi) from Krakatau, Anjer was a little farther, and the harbor at Telok Betong was 70 kilometers (45 mi) away: seemingly safe distances. These and other towns along the coasts were soon to be totally destroyed.

By 1 PM on August 26, loud explosions were occurring at intervals of about ten minutes, and a few hours later their intensity grew, rattling windows as far as Jakarta, 160 kilometers away. A British ship 120 kilometers distant took a sextant bearing on the top of Krakatau's ash cloud, and the captain noted in his log that the cloud now reached to an altitude of 25 kilometers (about 3 times the typical flight altitude of a modern commercial aircraft). More than fifty ships were in the Straits of Sunda at the time, enveloped in a midday darkness punctuated by lightning and the eery glow of Saint Elmo's fire in their masts and rigging. A continual rain of hot ash fell on the ships, and the crews choked in the sulfurous fumes as they struggled to shovel off the decks.

The climax came the next morning, August 27, with four monstrous explosions at 5:30, 6:44, 10:02, and 10:52 AM, times that were unexpectedly recorded on a pressure gauge chart at the Jakarta gasworks. These explosions were heard over vast distances: People were awakened in southern Australia, as far as 3,224 kilometers from the eruption, while at Diego Garcia, 3,647 kilometers to the west, the blasts were mistaken for cannon fire. Distant rumbles were heard over one-thirteenth of the earth's surface. Meanwhile, the inaudible lower-frequency components of these sound waves were detected on barometric recordings all over the world. Over the next five days, a series of seven equally spaced and particularly large fluctuations in atmospheric pressure were recorded by barographs in Bogota, Colombia, precisely on the opposite side of the globe from Krakatau. So violent were Krakatau's climactic explosions that they actually set the earth's entire atmosphere ringing like a bell. During these furious explosions, Krakatau's ash plumes ballooned upward to an altitude of at least 50 kilometers (31 mi) which defines the very top of the stratosphere.

Meanwhile, a huge *nuees ardentes* surged out to the north, raced across 30 kilometers (20 mi) of sea, and billowed *up* the mountain slopes of coastal Sumatra, where it incinerated several villages and up to 3,000 humans. Although the mats of floating pumice in the straits may have contributed to the great striking range of this pyroclastic flow (by providing an insulated surface for it to travel on), it nevertheless must have been extremely hot at its source if it still did such damage after traveling more than 30 kilometers. A Dutch woman, who with her children had taken shelter in a mountainside hut at an elevation of 200 meters (650 ft), survived the event only with great difficulty and with serious burns; in her later account, she reported that ash had burst through the wooden floor, consistent with what one would expect from a *nuees ardentes* deflected upward by the sloping terrain. Servants who had remained outside that hut were killed immediately.

Most devastating, however, were the tsunamis. Relatively small tsunamis struck nearby shores in the afternoon and evening of August 26, and again around 7 AM and 9 AM the following morning. Then at about 10 AM on August 27, roughly coinciding with Krakatau's major explosion, a series of monster waves innundated all the coasts in the region. The town of Telok Betong, situated at the mouth of a long, funnel-shaped bay that opened toward the volcano (Fig. 7.7), was completely swept away by a series of four great waves as high as 30 meters (over 100 ft). One ship succeeded in steaming out of Telok Betong's harbor and directly into the tsunami, which the captain described as resembling a series of mountains rising out of the sea. Another ship, the *Berouw*, was carried completely over the town and deposited in a forest 3 kilometers (2 mi) inland and 10 meters (33 ft) above sea level, coming to rest with her propellers pointing into the air. There the ship remained as late as 1979, when it was cut up for scrap. Dozens of other towns and villages and all their human occupants were obliterated by these great waves, which in places may have reached heights of 40 meters (130 ft). The tsunamis took

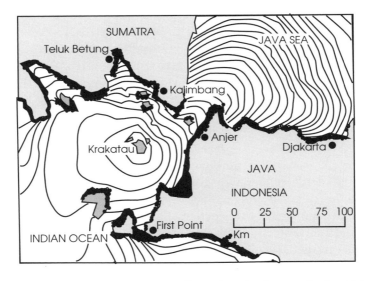

Figure 7.7. Krakatau and the surrounding region, showing areas hardest hit by the tsunamis of 1883. The expanding wave front is shown at intervals of about 5 minutes.

only eleven hours to reach Honolulu, and they were eventually recorded on tide gauges in all of the world's oceans.

The most destructive of the tsunamis seems to have resulted from the collapse of the volcano's walls into the magma chamber that had been emptied during the final great series of explosions. The newly formed undersea crater could not stand empty for long, and the seas rolled in. In fact, a ship that had been anchored just north of the Sunda Straits reported a tremendous southward current immediately prior to the tsunami. This suggests that the tsunami began with a wave trough rather than a wave crest, and that many coastal residents might have been able to escape if they'd noticed and recognized this. Unfortunately, there was very little light to see by, for the sky was completely black except for the flashes of electrical activity (which one always finds in turbulent clouds but which are intensified by an increased density of airborne particulates). Moreover, the noise level was probably too great to allow warning shouts to be heard, even if anyone did notice that the sea had receded from the shoreline. Of course, in those villages where there was no nearby higher ground to escape to, even as much as a half hour of advance warning would have been of little help. These conditions all surely conspired to allow those Indonesian waters to swallow up more than 30,000 human lives in the midday darkness of August 27, 1883.

The eruption tapered off quickly that afternoon, and by a little after midnight it was all over. The next morning, the Straits of Sunda were clogged with floating pumice as much 1 meter thick and thousands of human corpses. One full year later, a giant mat of this gruesome legacy of the eruption, pumice mixed with human skeletal remains, washed ashore in eastern Africa. It had floated a distance of 7,200 kilometers (4,500 mi).

Airborne ash from Krakatau's final explosions completely encircled the globe within two weeks, and for the next three years dramatic sunsets were observed everywhere on the planet. Although the volume of ash ejected into the stratosphere was not sufficient to affect global weather, the event did provide scientists with the first direct evidence of the circulation of upper atmospheric airstreams and led to a deeper understanding of what had happened after the larger explosion of Mt. Tambora, sixty-eight years earlier.[10]

Krakatau is not extinct. Eruptions in the undersea caldera continued to occur from time to time in the years after 1883, and in January of 1928 a new volcanic cone emerged from the sea (promptly named "Anak Krakatau," or "child of Krakatau"). Today, the caldera created in the spectacular eruption of a century ago is largely filled in with newly deposited volcanic debris, and as recently as the summer of 1995 the volcanic cone again spewed ash into the sky. Future generations will undoubtedly hear more from Krakatau.

The Skies Sometimes Fall

Large volcanic explosions are relatively rare, but when they do occur, they catch everyone's attention because of their awesome destructive power. Yet there is another category of natural disaster that is potentially a great deal more disastrous than even an exploding volcano: a large asteroid impact. This doesn't happen nearly as often as even the rare Tambora-sized eruption, and in fact there has not been a major event of this type since the first humans began to write. Nevertheless, we do know that when a large asteroid does hit our planet, as many have in the geological past, the effects are very similar to those of a major volcanic eruption: ashfalls, fires, tsunamis, atmospheric shock waves, and meteorological upheaval. A large-enough impact can do even worse: It can drive entire species to extinction and can permanently alter the course of biological evolution. Anthropocentric biases aside, there is no natural agent capable of creating greater destruction on our planet than a large asteroid that enters a collision course with Planet Earth. Although it's been a long time since the last big impact (around 50,000 years, if one defines a 1.2-kilometer crater as "big"), this does not mean that modern humans are immune from such a threat.

In the early morning of June 30, 1908, a giant fireball screamed across the sky over northern Siberia. Near the remote village of Tunguska it exploded so violently that it flattened 2,000 square kilometers (800 mi^2) of forest and set off an atmospheric shock wave that twice circled the planet. The intensity of the explosion has since been calculated as the equivalent of 20 million tons of TNT. Because the site is so remote, there were few if any human casualties, and no scientific investigation took place until a full 21 years later. Although no evidence of an impact crater could be found, the millions of fallen trees still lay, pointing away from the direction of the apparent center of the blast. Of the various hypotheses proposed, the most likely seems to be that a stony meteor, around 60 meters (200 ft) in diameter and weighing a few hundred thousand tons, crossed our planet's orbit and disintegrated explosively as it crashed into Earth's atmosphere at a speed of more than 100,000 kilometers per hour.[11] After the shock wave knocked down the trees, the remnants of the meteor apparently settled quietly to the earth as a gentle dust.

Our solar system is mostly empty, our own planet a relatively tiny speck that orbits the sun at a speed of about 107,000 kilometers per hour (67,000 mi/h). Earth regularly slams into uncountable numbers of tiny sand-grain-sized particles that flash briefly in the upper atmosphere as they disinte-grate. These are the "shooting stars," or *meteors,* visible to the naked (and patient) eye from the surface of the planet on virtually any clear night. Occa-sionally a meteor is a bit larger, and if it is composed of iron rather than stone, it stands a chance of reaching Earth's surface intact, after being slowed by atmospheric drag during its brief descent (which takes only a few seconds). It is then referred to as a *meteorite;* examples can be seen in the geology exhibits at many museums.

The most recent documented case of an extraterrestrial object striking a human dates back to 1954; a woman in Sylacauga, Alabama, resting on the sofa in her living room, was struck and bruised by a 5-kilogram (11-lb) meteorite that smashed through her roof and ricocheted off a radio. More curious, however, are the two most recent cases of meteorites striking houses (but not people) in the United States. In April of 1971, a meteorite caused minor damage to a house in Wethersfield, Connecticut. Eleven years later, on November 8, 1982, another meteorite damaged a second house *in the same town!* In an amazing coincidence of two random events, the two impacts were only about 1 kilometer apart.[12]

These, however, were far from catastrophic events. The probability of a catastrophic meteoric collision, fortunately, is quite small, at least for any given year. In taking a census of interplanetary objects in the inner solar sys-tem, astronomers find fewer and fewer objects as they move to larger sizes. Any object larger than 1 kilometer or so (yet considerably smaller than a planet) is referred to as an *asteroid.* The motions of asteroids tend to be much more regular and stable than those of sand-grain-sized particles, so once a large asteroid is discovered, its orbit around the Sun can usually be pre-dicted reliably for many centuries to come. At present, at least 300 objects greater than 1,000 meters (3,000 ft) in diameter are known to have orbits that intersect Earth's,[13] and about 30 new objects of this size are being discov-ered and added to the inventory each year. Fortunately, none of these known objects is likely to collide with Earth anytime soon. Unfortunately, this ongoing search is estimated to be only around 8% complete, so we remain far more ignorant than knowledgeable about the potential risks. Asteroids are extremely difficult to detect because their surfaces reflect so little light, and we usually discover them only when they eclipse a series of distant stars over a period of days or weeks. In fact, in 1989, a 1,000-meter asteroid was discovered only *after* it had crossed Earth's orbit, at a spot where Earth had been only six hours earlier. In May of 1996, another asteroid of roughly the same size was first discovered only four days before it sped across Earth's orbit, ultimately missing our planet by four hours. To many scientists these were uncomfortably near misses. These incidents have, however, raised the level of scientific interest in searching for earth-crossing asteroids.

Statistical analyses suggest that Planet Earth can expect a collision with an asteroid of 1,000-meter diameter or larger about every 250,000 years, on average. For smaller objects, which are considerably more numerous, we expect the collision probability to be quite a bit higher. The graph in Figure 7.8 shows the approximate rates of collision between Earth and extraterrestrial objects of different sizes. Although a graph like this is based on quite a few theoretical assumptions, an iron meteorite roughly 50 meters (160 ft) in diameter did strike in Arizona just 50,000 years ago, which is extremely recent in geological time. The crater, which remains quite prominent today, measures 1.2 kilometers (4,000 ft) in diameter, or about 25 times the diameter of the speeding hunk of iron that created it. The energy

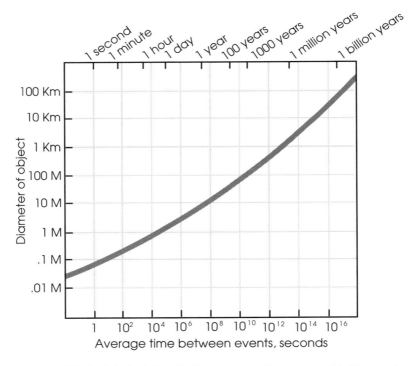

Figure 7.8. Statistical relationship between sizes of extraterrestrial objects and their frequencies of collision with Planet Earth. Impact craters are typically about 25 times the diameter of the impacting object.

released in that impact was the equivalent of at least a 4-megaton nuclear explosion.[14]

If we are willing to broaden our conceptual horizon to a time line that extends back a few hundred million years, we can re-create even more dramatic asteroid impacts from the past. Table 7.2 lists the ancient craters that have been discovered with diameters of 40 kilometers or greater.[15] Undoubtedly there are others this big whose features have been obliterated by the relentless forces of erosion, or that are hidden beneath the seas that cover 70% of our planet. Incomplete data notwithstanding, there does seem to be considerable evidence that these impacts aren't freak events, but rather that they occur with some statistical regularity. It should be noted that even the smallest craters listed in Table 7.2, 40 kilometers in diameter, had to have been created by asteroids with diameters of at least 1.6 kilometers, or roughly 1 mile. Big objects like this *do* seem to strike Earth from time to time, at least when we speak of geological time spans.

Some readers may remain skeptical that such events are a real threat to humanity. The Tunguska meteor, after all, never did hit the ground, and that is our single significant incident of this type within historical times. Are

Table 7.2. *The known impact craters with diameters of 40 km or larger*

Crater name	Location	Approximate diameter (km)
Vredefort	South Africa	300
Sudbury	Ontario, Can.	250
Chicxulub	Yucatan, Mexico	170
Manicouagan	Quebec, Can.	100
Popigai	Russia	100
Acraman	South Australia	90
Chesapeake	Chesapeake Bay, USA	85
Puchezh-Katunki	Russia	80
Kara	Russia	65
Beaverhead	Montana, USA	60
Tookoonooka	Queensland, Australia	55
Charlevoix	Quebec, Can.	54
Siljan	Sweden	52
Kara-Kul	Tajikistan	52
Montagnais	Nova Scotia, Can.	45
Araguainha Dome	Brazil	40
St. Martin	Manitoba, Can.	40

major collisions with asteroids really a credible menace? Let me describe two other events that may give the skeptical thinker pause to wonder.

On June 25, 1178, five British monks witnessed an event that in retrospect seems to have been an asteroid striking the Moon. The Moon was in a thin-crescent phase, and near the upper horn of the crescent the monks saw what they described as a "flaming torch" that ejected a dramatic display of fire and sparks. Then the entire crescent went dark, as one might expect as the dust settled slowly in the weak gravity and blocked the sunlight that normally reflects from the Moon's surface. The reported position of this apparition coincides closely with the lunar crater we now call Bruno. Alone, however, this curious account was never compelling enough to convince most scientists, and until about 25 years ago, no one paid much attention to it. Then, in the early 1970s, several instrument packages were set up on the surface of the moon, and the data they beamed back showed that the entire Moon is vibrating like a huge bell with a period of about three years. This is remarkably consistent with what one would expect a mere 800 years after a major asteroid impact in the general region of the monks' documented observations.[16] If we apply Occam's Razor, we need to seriously entertain the possibility that a large asteroid collided with the Moon, in an impact so energetic that it was visible to the naked eye at a distance of some 390,000 kilometers (240,000 mi) just a little over 800 years ago.

More recently and even more dramatically, during the week of July 16 through 22 of 1994, a series of more than twenty large comet fragments crashed into the planet Jupiter.[17] Several of these chunks of ice and stone were as large as 5 kilometers in diameter, and their impacts ejected great plumes of vaporized material high above the Jovian atmosphere.[18] This tremendous series of explosions, widely observed by telescope, underscored the fact that violent collisions between large objects in space are more than mere theoretical possibilities. They not only *can* happen, but in fact they *do* happen from time to time.

But just how serious is the risk to the humanity that populates Earth? Statisticians usually assess the risk per person by dividing the potentially affected population by the product of the number of events and the average expected number of deaths per event, adjusted to a specified time interval. This sounds like a mouthful, but it becomes easier to understand by comparing some results. In Table 7.3, I list the risk of an individual's dying during a 50-year period due to a few of life's many hazards. On this basis, we see that the risk of death from an asteroid impact lies in the same range of probabilities as some other well-known risks: death caused by a hurricane or an airplane crash, for instance. These numbers reflect the fact that even though

Table 7.3. *Average risk of death to an individual over a fifty-year period, as estimated from historical frequencies and current populations*

Worldwide: Risk of death from	
Volcanic eruption	1 in 30,000
Asteroid impact	1 in 20,000
United States only: Risk of death from	
Earthquake	1 in 200,000
Lightning	1 in 130,000
Tornado	1 in 50,000
Hurricane	1 in 25,000
Airplane crash	1 in 20,000
Electrocution	1 in 5,000
Auto accident	1 in 100

major asteroid impacts are unlikely in a given year, when they do occur they have the potential to generate destruction of enormous proportions.

Many of the devastating effects we've already associated with volcanoes could be produced just as effectively by a collision between our planet and an asteroid. Approaching at a typical 107,000 kilometers per hour, a large asteroid will plow through Earth's 150 kilometers of atmosphere in just 5 seconds, and (unlike the much lighter and ubiquitous micro-meteors) it will decelerate very little before smashing into the surface. If a city happens to lie at ground zero, a 1-kilometer asteroid will instantly obliterate most evidence of human life or activity. Striking land or shallow water, it will throw huge masses of dirt and debris high into the atmosphere, disrupting global climate for many years, if not decades. A major impact in an ocean will set off a tsunami of monumental proportions that will scour most of the world's coastlines of all structures within many kilometers of shore.

This much is probably obvious. What is less obvious is the possibility of two additional effects: induced earthquakes and volcanoes at the antipode of the impact site, and an induced meteorological event called a "hypercane."

Any major impact will send a shock wave into Earth, and this wave will reflect and refract in complicated ways as it passes through the crust, the layers of the mantle, and the inner and outer core. Because of the planet's spherical symmetry, however, the energy of such a shock wave will converge toward a general region centered on the opposite side of the globe from the impact point. If the crustal region at this antipode already happens to be under tectonic stress, the converging energy may push it beyond its threshold, setting off secondary earthquakes and perhaps even volcanic eruptions. The seismological record is very clear that earthquakes are sometimes set off by other earthquakes, and the geological record seems to support speculations that some prehistoric lava flows may have been induced by asteroid impacts. Mathematical models based on such available observational evidence, fuzzy though the numbers may be, tell us that this is not a farfetched scenario: An asteroid impact on one side of the globe might easily trigger earthquakes and even volcanic eruptions on the opposite side of the planet.

Should a large asteroid plummet into one of the world's oceans, another of its effects would be to heat the water at the impact site. We already know that warm ocean water induces atmospheric instabilities that lead to tropical storms and hurricanes. The more localized, but hotter, water at the site of an asteroid impact is likely to induce a related meteorological phenomenon: a hurricane that is relatively compact for a hurricane but extremely intense: a *hypercane*. Although the current mathematical models for hypercanes remain riddled with uncertainties, there can be no doubt that any major asteroid impact in an ocean would generate violent meteorological effects that would persist long after the immediate event and would do great damage far from the impact site.

Serious proposals have been made, some even drawing attention in U.S. Congressional hearings, to protect Planet Earth from collisions with asteroids by reallocating the world's useless and expensive nuclear arsenals for this purpose. Unfortunately, this idea is much more easily proposed than carried out, and numerous technical and political problems would need to be solved first. It is not even clear that one could reliably detect an approaching asteroid while it was still far enough away to do something about its orbit. Asteroids are awfully tiny objects against the background of space; only if they happen to hit us will they suddenly seem big.

The ultimate irony is this: We humans wouldn't be here to ponder this issue if it weren't for a cataclysmic asteroid impact that occurred about 65 million years ago. At that remote time, the dinosaurs were the dominant life form on the planet, as they had been for at least the previous 100 million years. (Compare this to the paltry 50,000 years or so that *homo sapiens* has existed as a species.) Dinosaurs were far from fragile creatures; they surely wouldn't have been around so long if they had not been superbly adapted to Earth's environment. The small, primitive mammals that coexisted with the dinosaurs stood far from the top of the food chain. Then suddenly, throughout the world, the dinosaurs disappeared from the fossil records, and the little mammals quickly diversified and grew to fill the empty ecological niches. We humans are among the descendants of small, furry, and prolific creatures whose original purpose in nature's scheme was to serve as dinosaur food.

The question of what happened to the dinosaurs arose immediately when their fossilized bones were first discovered in the early 1800s, but only a little more than a decade ago did a credible scientific theory emerge.

Asteroid Impacts and Mass Extinctions

Between the late 1950s and early 1960s, a physician named Immanuel Velikovsky published a series of three books that were received enthusiastically by millions of readers worldwide but were subjected to widespread ridicule in the scientific community. His first, *Worlds in Collision*, replete with long and scholarly footnotes citing ancient manuscripts from cultures around the globe, proposed the thesis that Earth had experienced a series of near-misses with a large comet within historical times. This comet was to have somehow been ejected by the planet Jupiter; then, as it made a number of passes through Earth's orbit, its gravitational effects serendipitously produced a whole series of miracles recorded in the Old Testament (e.g., the collapse of the walls of Jericho, the Sun stopping in the sky, the parting of the Red Sea, the manna falling from heaven during the Exodus, and so on). Velikovsky's comet didn't disappear back into deep space but was deflected by Earth's gravity to fall into a stable, nearly circular orbit around the Sun. Eventually, according to Velikovsky, this body came to be named Venus.

Now all of this is quite demonstrably a bunch of nonsense, for it not only defies the principles of orbital mechanics, but we also know quite well through our space probes and spectrographic analyses that the composition of Venus has no similarity whatsoever to the composition of Jupiter. Further, the historians assure us that Venus was well known to the ancients before the time of the Exodus, and that it tracked along its present orbit even in that remote time. Yet Velikovsky's books made fascinating reading and received a great deal of favorable publicity when he wrote them. He visited quite a few university campuses in the late 1960s, and (to the chagrin of many science faculty) his lectures were always well attended by students.

Despite his intellectual floggings by the scientific community, Velikovsky's ideas continued to attract followers whose numbers diminished only very gradually over the next two decades. As a result, the scientific climate during this period was far from supportive of any researcher who might be audacious enough to propose an extraterrestrial origin for any past global catastrophe. In fact, the very concept of global catastrophe was pretty much taboo as a subject of serious scientific inquiry.

Then, around 1980, Luis Alvarez and his coworkers sought to explain a curious scientific observation.[19] At sites around the world, chemical analyses kept showing an unexpectedly high concentration of the rare element iridium in geological samples taken from the narrow boundary between the Cretaceous and Tertiary periods of prehistory. This boundary layer, which was deposited about 65 million years ago, also marks the disappearance of the dinosaurs from the fossil record. Below (and as much as 100 million years before), one can find dinosaur fossils, but above, never. Was it just a fluke that this same microscopic boundary that dates the extinction of the dinosaurs also contains elevated levels of iridium?

The only other places one finds high concentrations of iridium are in discarded catalytic converters and in certain metallic meteorites. Alvarez and his research team theorized that a large metallic asteroid struck Earth 65 million years ago and ejected great quantities of particulates into the upper atmosphere, where they blocked the Sun. With its radiation balance upset, Earth quickly cooled to the point where it was no longer hospitable to giant cold-blooded inhabitants. A few years of dark skies and cold would have been more than sufficient to guarantee the extinction of the dinosaurs throughout the world. Eventually the iridium-laden dust settled uniformly over the planet's surface, and over the bones of the dinosaur carcasses, providing us with the chemical evidence to reconstruct the event today.

When first proposed, this was far from a popular theory (though no scientist would dispute that it was more scientific than Velikovsky's). Researchers traveled all over the world to test for iridium at the Cretaceous–Tertiary (K–T) boundary, hoping to find samples that would destroy Alvarez's theory. Instead they continued to find even more examples of elevated iridium concentrations that corresponded very neatly to the time the dinosaurs disappeared.

Yet despite this additional corroborative evidence, the impact theory of dinosaur extinction still lacked a "smoking gun." Where is the crater? To account for global effects on a scale that would kill off thousands of species, the crater had to be very big, on the order of 250 kilometers in diameter, corresponding to an asteroid at least 10 kilometers in diameter. Even after 65 million years of erosion, some evidence of such a dramatic surface feature should remain. Here all the results are not quite in yet, but it does seem that a recently discovered crater (Chicxulub), having the right approximate age and nearly the right size, lies mostly underwater at the southwestern shore of the Yucatan Peninsula in Mexico. It's still not time to give out cigars, but scientists are finally beginning to climb on the asteroid impact bandwagon, and many are seriously looking for additional corroborating evidence. We may someday conclude, as unambiguously as possible, that the dinosaurs were indeed driven to extinction by the collision of a large asteroid with Planet Earth.

The idea that major natural disasters of the past may have had extraterrestrial origins has begun to enter the mainstream of scientific thought. The fossil records reveal fairly clearly that there have been five major global extinctions: the first some 450 million years ago, then 350 million, 225 million, 190 million, and finally 65 million years ago. In just the last few years, serious scientific hypotheses have been proposed which link each of these known biological catastrophes to astronomical events such as asteroid impacts, supernova explosions, and the like. Although at present none of these theories is as compelling as the Alvarez theory for the Cretaceous–Tertiary extinction, what *has* happened is that today's scientists are considerably more open to the possibility that our planet may be vulnerable to future global disasters of extraterrestrial origin. Expectations for the long-term survival of the human species can claim no preferred status in the Cosmic order.

Until fairly recently, scientists figured that no natural event could possibly threaten all life on Earth for at least another 5 billion years, when (according to astrophysical calculations) our Sun will start to swell into an enormous red giant, the oceans will evaporate, and our planet will be reduced to a cinder. Now it appears that global catastrophes with the potential of rendering Earth lifeless may occur as often as every 100 million years on average, and that impacts sufficient to destroy most human civilization on the planet may happen once every 300,000 to 1 million years. Worse, unlike the predictable burning out of the sun, such global catastrophes may occur with little or no warning. Is this something to lose sleep over? I, for one, will be more comfortable if I know *someone* knowledgeable is losing sleep over it.

Notes

1   Contemporary accounts of the disaster appear in A. Heilprin, Mont Pelée in its might, *McClure's Magazine,* Oct. 1902, 359-68, and C. Morris, *The destruction of St. Pierre and St. Vincent and the World's Greatest Disasters . . .* (Philadelphia: American Book and Bible House, 1902). A more recent article is L. Thomas, Prelude to doomsday, *American Heritage,* Aug. 1961, 4–9 and 94–101.

2 S. Chretien, & R. Brousse, Events preceding the great eruption of 8 May 1902 at Mount Pelée, Martinique, *Journal of Volcanological Geothermal Research*, 38 (1989), 67–75.

3 One eyewitness, Roger Arnoux, a member of the French Astronomical Society, reported that he saw the cloud travel some 8 kilometers in "not more than three seconds." If so, its speed had to be 8 times the speed of sound. This does not seem to be physically possible and underscores the problem of taking eyewitness accounts too literally.

4 Comparative studies of the two events are meager. One of the few is A. G. MacGregor, Eruptive mechanisms: Mt. Pelée, the Soufrière of St. Vincent and the Valley of Ten Thousand Smokes, *Bulletin Volcanologique*, 12 (1951), 49–74.

5 T. Simkin & L. Siebert, *Volcanoes of the world* (Washington, D.C.: Smithsonian, 1994).

6 Sir Thomas Stamford Raffles, *History of Java* (London, 1817; reprint, Oxford: Oxford University Press, 1965).

7 R. B. Stothers, The great Tambora eruption of 1815 and its aftermath, *Science*, 224 (1984), 1191–8.

8 H. Stommel & E. Stommel, *Volcano weather: The story of 1816, the year without a summer* (Newport, R.I.: Seven Seas, 1983). Among the several articles the same authors published on the same subject: The story of 1816, the year without a summer, *New Scientist*, 102 (1984), 45–53.

9 T. Simkin & R. S. Fiske, *Krakatau* (Washington, D.C.: Smithsonian, 1983).

10 P. Francis & S. Self, The eruption of Krakatau, *Scientific American*, Nov. 1983, 172–87.

11 F. J. Whipple, The great Siberian meteor, *Quarterly Journal of the Royal Meteorological Society*, 56 (1930), 287-394. For a more recent article, see R. Ganapathy, The Tunguska explosion of 1908: Discovery of meteoritic debris near the explosion site and at the South Pole, *Science*, 220 (1983), 1158–61.

12 B. Berman, Struck by a meteor, *Discover*, Aug. 1991, 28.

13 E. M. Shoemaker, R. F. Wolfe & C. S. Shoemaker, Asteroid and comet flux in the neighborhood of Earth, in V. L. Sharpton & P. D. Ward, eds., *Global catastrophes in Earth history*, Geological Society of America Special Paper no. 247 (Boulder, Co.: Geological Society of America, 1990).

14 H. C. Urey, Cometary collisions in geological periods, *Nature*, 242 (1973), 32.

15 R. Grieve, The record of terrestrial impact cratering, *GSA Today*, Oct. 1995.

16 J. D. Mulholland & O. Calame, Lunar crater Giordano Bruno: A.D. 1178 impact observations consistent with laser ranging results, *Science*, 199 (1978), 875.

17 D. J. Eicher, Death of a comet, *Astronomy* 22 (10) (Oct. 1994), 40–45.

[18] The observation of discharge of material into space from Jupiter is particularly impressive, given that Jupiter's gravity is stronger than that of Earth by at least a factor of 3.

[19] L. W. Alvarez, W. Alvarez, F. Asaro et al., Extraterrestrial cause for the Cretaceous–Tertiary extinction, *Science*, 208 (1984), 1095–1108. See also W. Alvarez & R. A. Muller, *Nature*, 308 (1984), 718–20, and W. Alvarez et al., *Nature*, 216 (1982), 886–8.

8

Deadly Winds

Dade County, Florida, 1992

Two weeks into August 1992, a tropical depression developed off the western coast of Africa and began tracking westward. Such rainstorms are common over tropical waters in late summer and early fall, they usually do drift toward the west, and most run their course within a few days. Yet they always bear watching, and in this instance the meteorologists studying their satellite images became modestly concerned as the wind speeds continued to grow. When sustained winds exceeded the 63 kilometer per hour (39 mi/h) level, standard protocol dictated that the event be assigned a name from a predetermined list: "Andrew," the *A* denoting the first full-blown tropical storm of the season.

Still, there didn't seem to be serious cause for alarm, for as Andrew approached the Caribbean, he veered toward the north and out of the tropics. At latitudes north of the Tropic of Cancer (23.5 °N), the prevailing winds usually shear the tops from the thunderclouds of a tropical storm, and this prevents the storm from growing further into a full-blown hurricane.[1] Andrew, however, didn't get this message, and at a latitude of nearly 26 degrees north, his sustained winds escalated beyond 119 kilometers per hour (74 mi/h), earning him the designation of a Category 1 hurricane.

By 11 PM on Saturday, August 23, after crossing more than 5,000 kilometers of open ocean, Andrew was 840 kilometers (520 mi) due east of Miami and traveling due west at 22 kilometers per hour (14 mi/h). Within the hurricane, winds were now gusting to 175 kilometers per hour (110 mi/h), which flirted with a Category 3 rating. If the storm didn't change its course, it would strike southern Florida in less than thirty-seven hours, and damage was certain to be extensive.

The paths of hurricanes, however, are notoriously unpredictable, and these violent storms have been known to make some wild twists and turns in thirty-seven hours. To give an evacuation order too soon is to risk evacuating the wrong areas, which is certain to undermine the credibility of other evacuation orders in years to come. On the other hand, a delay in ordering an evacuation incurs grave risks. Responsible authorities need to avoid having great throngs of people evacuating at night or jamming the freeways after heavy rains and flooding have already begun. In fact, during the "Labor Day Hurricane" of 1935, a late and unenthusiastic evacuation order stranded thousands of people in the Florida Keys as the hurricane arrived. Four hundred eight people died in that disaster.

Miami had suffered direct strikes by hurricanes in 1926 and 1928, and several hundred people died in those two events. The next direct hit came in 1950, but that storm caused relatively minor damage and few casualties. Now, forty-two years later, the Miami metropolitan area had mushroomed, the majority of its buildings had never been subjected to high winds, and most of the area's residents (nearly 2 million in Dade County alone) had never experienced a serious tropical storm. The situation confronting the Civil Defense decisionmakers was fraught with uncertainties on multiple levels. To their credit, all of the authorities, news teams, and public service personnel did an outstanding job. When the winds and rain finally subsided, having claimed a huge $30 billion in property damage and upsetting the lives of more than 3 million people, Andrew's human toll in Florida stood at only 43.

On Sunday morning, August 23, the Florida newspapers announced the oncoming hurricane with blaring headlines that few could miss.[2] Evacuation orders for low-lying areas were issued before noon on every radio and television station, and more than 1 million people evacuated their homes and rolled out their sleeping bags in designated hurricane shelters. Paramedics and police rounded up the homeless and transported them to public shelters. Meanwhile, every piece of plywood in every lumberyard quickly sold out, as people scrambled for materials to use in boarding up their windows and doors. Huge traffic jams piled up on all roads leading from the coast. This mass pandemonium, however, lasted only a few hours. By late afternoon, when the skies grew overcast and a gentle rain started to fall, the traffic had already thinned out. The last of the businesses closed up, and hospitals moved all of their patients to interior corridors. Engineers tested emergency lighting systems. The winds picked up, and millions huddled with loved ones to spend the most anxious night of their lives.

Although Andrew had been expected to strike Miami head-on, its center in fact passed 24 kilometers (15 mi) to the south and arrived a few hours earlier than anticipated, making landfall around 4:52 AM on Monday, August 24. At that time the hurricane was plowing forward at a fairly rapid 40 kilometers per hour (25 mi/h). We don't know the peak wind speed, because the winds destroyed most of the instruments at the National Hurricane Center in Miami (even though that facility was well outside the main swath of the storm). All evidence suggests, however, that this was a Category 5 hurricane (the highest on today's scale), with sustained winds in excess of 250 kilometers per hour (155 mi/h). Some localized gusts may even have been as high as 300 kilometers per hour (200 mi/h), a velocity more typical of tornadoes than hurricanes.[3]

Because of its relatively high forward speed and compact size, Andrew became Dade County history after only around seven hours. By noon, most of the evacuees were out in the residual drizzle trying to make their way through fallen trees and power lines to the remains of their homes. In too many cases, there was very little left standing (Fig. 8.1). The major devastation lay in a band 30 kilometers (20 mi) wide that ran through the residential communities southwest of Miami, and the entire town of Homestead lay virtually flattened. Although it is extremely difficult and expensive to build a structure that can survive unscathed in the highest gusts of a hurricane like Andrew, later investigations revealed that many of the existing hurricane construction codes had been laxly enforced, and there is now little doubt

Figure 8.1. Destructuion in Homestead, Fla., in the aftermath of Hurricane Andrew, 1992. (Photo courtesy Tim Marshall.)

that building code violations seriously aggravated the effects of the disaster. The resulting avalanche of insurance claims forced at least six Florida-based insurance companies into bankruptcy, and litigation against building contractors tied up the Florida courts for several years. Many of those who lost homes and possessions never did receive compensation for more than a fraction of their losses.

Although Andrew lost some of his punch as he crossed the land mass of southern Florida, he received a new infusion of energy as he entered the warm waters of the Gulf of Mexico and turned toward the north. New Orleans, which is mostly at or below sea level, prepared for a direct hit and major flooding. Fortunately for that city, Andrew unexpectedly veered to the west in the last hours before his second landfall, and the human misfortunes fell elsewhere.

Around 11 PM on August 25, the hurricane struck shore near Morgan City, Louisiana, a town of 15,000 people, most of whom had been wise enough to evacuate. At that time, sustained winds were being clocked at 225 kilometers per hour (140 mi/h), and the hurricane's forward progress had slowed to 21 kilometers per hour (13 mi/h). This low storm speed gave the high winds enough time to raise a huge storm surge that flooded a great section of coastline. Fifteen people died in Louisiana, and Andrew added $2 billion to his property damage price tag. Offshore, the barrier island Grand Isle disappeared under water completely during the storm, just as Galveston Island had been swallowed by the sea during its great disaster of the year 1900.

The hurricane's fury dissipated quickly as it drove itself inland; by 1 PM on August 26 it was downgraded to a tropical storm with wind speeds of 96 kilometers per hour (60 mi/h), and twelve hours later its winds dropped to a modest 55 kilometers per hour (35 mi/h) as it passed through Georgia. At this point Andrew was no longer capable of doing anything more serious than dumping a lot of rain.

But in southern Florida, at least 160,000 people were homeless (some sources claim as many as 300,000), 80,000 dwellings lay flattened, another 55,000 houses were seriously damaged and only partly habitable, a million people were without power, and food and water were scarce. Relief efforts were hampered for many days by the devastation of the infrastructure: impassible streets, downed power lines, and interruptions in phone service. To make matters worse, it rained for most of the following week, and many of the donations of clothing that were trucked in from around the country were soaked and ruined when they were unloaded and stacked in the open (for lack of any remaining roofs to put the materials under). Looters moved into the devastated areas, and National Guard troops were deployed in response.[4] Many losses were uninsured or underinsured, and even many of the insurance companies that did survive were brought to their knees. Much of what Andrew destroyed has still not been rebuilt.

Some writers contend, on the basis of the dollar value of the destruction, that the Northridge, California, earthquake of January 17, 1994, was a greater natural disaster than Hurricane Andrew. If dollars are the prime criterion, this indeed may be true: some $35 billion in earthquake damage in the Los Angeles area in 1994, compared to a roughly $32 billion bill, mainly in Florida, due to Andrew in 1992. Hurricanes do not cause freeways to collapse nor water and gas lines to rupture. Complicating any comparison is the fact that dollars of private property loss may be more a reflection of local real estate market conditions than of comparable levels of destruction. There can be no argument, however, that Andrew did destroy many more houses and dislocate many more people than did the Northridge earthquake. It was only in February of 1995, two and one-half years after the hurricane, that the last of the victims housed in temporary trailers were moved to permanent living quarters. A full four years after the event, more than 20% of the damaged buildings in Florida still had not been repaired or replaced. If the criterion is total human impact, Hurricane Andrew must be ranked undisputably as the most destructive natural disaster in U.S. history.

Dynamics of the Atmosphere

We live at the bottom of an ocean of air, on a rapidly rotating planet whose surface is heated unevenly by the Sun. Earth's surface radiates heat energy back into space, different parts radiating at different rates. Because the total incoming solar energy balances the total outgoing radiation energy, the planet's overall average temperature stays fairly constant (although there is evidence that it may currently be warming very slightly). Locally, however, there can be wild variations in the energy balance. It is these local imbalances, coupled with Earth's rotation, that drive our weather.

It's convenient to describe the atmosphere as a series of concentric shells that surround the globe like the layers of an onion. The lowest of these atmospheric shells is called the *troposphere;* it extends to an altitude of about 11 kilometers (7 mi), or just a few kilometers higher than the tallest mountains. The bottom of the troposphere is at about the same temperature as Earth's surface; at its top, however, it averages   55 °C (70 °F). Most clouds, and all weather phenomena, are confined to the troposphere. Above this region is a thinner layer called the *tropopause*, which extends to an altitude of about 16 kilometers (10 mi). At this point the atmosphere has a very low density, although it can still support jet aircraft in flight. The next layer is the *stratosphere*, which extends to 50 kilometers (30 mi) above Earth's surface but has a fairly constant temperature throughout. Above this, we have the *mesosphere*, the *thermosphere*, and finally the *exosphere*, which gradually trails off into space.

On the scale of the planet, all weather is confined to an extremely thin atmospheric blanket. Although Earth is roughly 6,400 kilometers in radius, the troposphere extends upward only an additional 11 kilometers or so, which amounts to just two-tenths of 1% of the distance to Earth's center. It should not surprise us, then, that Earth's weather is strongly influenced by the interactions between the troposphere and Earth's surface. These interactions take many forms, but we can summarize the principal phenomena as follows:

1. Uneven heat transfer from Earth's surface to the atmosphere. Land masses warm up and cool off much more quickly than do bodies of water.
2. Atmospheric convection. Warm air rises, with cooler air dropping to take its place.
3. Evaporation. Warm water evaporates more quickly than cool water, and it evaporates even more rapidly in high winds.
4. Expansion cooling. Air expands as its pressure decreases, and this results in a drop in temperature.
5. Condensation. Water vapor condenses from the atmosphere when the temperature drops below a threshold level (which depends on the pressure and humidity).
6. Condensation warming. Water vapor gives up heat when it condenses to form clouds and rain.
7. Pressure-induced flow. Air masses generally flow from regions of high atmospheric pressure toward regions of lower atmospheric pressure.
8. The Coriolis effect. Earth's rotation causes large, moving air masses to follow curved rather than straight-line paths. This results in large-scale counterclockwise circulations in the Northern Hemisphere and clockwise rotations in the Southern Hemisphere.
9. Solar radiation blocking. Clouds cast shadows, which affect the balance between incoming sunlight and the heat radiated from Earth's surface.

10. Day-night cycles. Local radiation balances vary over the course of any twenty-four hour period.
11. Turbulence. Air seldom flows smoothly from place to place but tends to produce eddies and swirls across a wide range of scales.
12. The Bernoulli effect. Atmospheric pressure drops as wind speed increases.

Most of these phenomena are probably already familiar to the reader in some form or another. What may be less apparent is that they are all intimately interconnected. The Sun cannot heat Earth's surface without setting up temperature differentials between land and water, which in turn causes air to rise at a greater rate in one place than another, which results in atmospheric pressure differences, which create winds, which swirl into small and large eddies and carry moisture from one place to another, in turn affecting temperatures and pressures, and so on.

There are obviously many variables that drive the behavior of the atmosphere, and the web of mathematics that relates them is far from simple. In some details, particularly those relating to turbulent flow, our current mathematical models of atmospheric dynamics are still quite incomplete. But beyond such gaps in our understanding, there may be a fundamentally more serious problem confronting any attempt to predict storms and weather: The phenomenological boxes we use to describe various elements of atmospheric behavior simply may not stack up to give the whole story. Our way of doing science is reductionistic, while the atmosphere behaves holistically. There may be important respects in which the reality is different from the sum of its perceived parts.

Will we ever succeed in predicting atmospheric behavior to any reasonable degree of accuracy weeks or months in advance? Is it even theoretically possible to do so? Although the scientific jury is still out on this question, current evidence suggests that Mother Nature may place intrinsic limits on the level of detail we humans might ever expect to achieve in predicting the behavior of complex systems like the atmosphere.

We will return to this point in more detail in Chapter 9. For now, let's look at some of the things we *do* seem to know.

Tropical Cyclones and Hurricanes

Internationally, *tropical cyclone* is the general term given to any large circulating mass of clouds and rain that develops over tropical waters. These potentially destructive weather patterns arise in the two bands of latitude ranging from about 7° to 25° north and south of the equator. Because the prevailing winds in the tropics drive tropical cyclones in the general direction of east to west, they are unlikely to strike west-facing coastlines. (The exceptions are the Pacific coasts of Mexico and Central America, which actually face southwest and are occasionally dealt a glancing blow by one of these storms.) Table 8.1 summarizes the classifications of tropical cyclones. When the circulating winds exceed 63 kilometers per hour (39 mi/h), the event is classified as a *tropical storm*, and at wind speeds that exceed 119 kilometers per hour (74 mi/h) it becomes a hurricane. A hurricane is the same meteorological phenomenon that is referred to as a "typhoon" in the western Pacific and as a "cyclone" in regions bordering the Indian Ocean. Of course, hurricanes themselves vary widely in intensity; Table 8.2 summarizes the Saffir–Simpson Scale for rating hurricanes in categories from 1 through 5.

In a typical year, around 100 tropical storms develop worldwide, two-thirds of them in the Northern Hemisphere. All of these storms spend most of their lives over water, and a majority of them never do make landfall. The

Table 8.1. *Classification of tropical cyclones*

Tropical disturbance	Rotary circulation is apparent aloft, although it may be slight or absent at the surface. No strong winds. Isobars (lines of equal atmospheric pressure) are discontinuous. A common phenomenon in the tropics.
Tropical depression	Some rotary circulation extends down to the surface. Winds do not exceed 63 km/h (39 mi/h). At least one isobar forms a continuous closed loop.
Tropical storm	Distinct rotary circulation over a range of altitudes. Wind speeds between 63 km/h and 119 km/h (39 mi/h to 74 mi/h). Closed isobars.
Hurricane	Strong and very pronounced rotary circulation. Sustained wind speeds exceed 119 km/h (74 mi/h). Closed isobars.

Note: Clouds and rain are present in all cases.

Table 8.2. *The Saffir–Simpson Scale of hurricane intensities*

Category	Damage	Wind speed (mi/h)	Barometric pressure (in.)	Surge (ft)
1	Minimal	74–95	28.94 or more	4–5
2	Moderate	96–110	28.50–28.91	6–8
3	Extensive	111–30	27.91–28.47	9–12
4	Extreme	131–55	27.17–27.88	13–18
5	Catastrophic	>155	27.17 or less	>18

Note: Normal barometric pressure at sea level is 1013.250 millibars, which is equivalent to 29.92126 inches on a mercury barometer, or a pressure of 14.69595 lb/in.^{2}. The lowest sea-level barometric pressure ever recorded was during Hurricane Gilbert in 1988: 887.9 millibars, or 26.22 inches of mercury.

15 or so that develop in the eastern Pacific usually blow out to sea before causing much trouble. The northern Indian Ocean generates about 12 a year, 8 of which may reach hurricane strength; occasionally one of these will wreak considerable devastation along the coastlines of India and/or Bangladesh. The western Pacific Ocean is the world's most prolific generator of tropical storms, averaging around 30 a year, some 20 of which may reach hurricane strength (although they are still called "typhoons"). Again, most of these storms blow themselves out over the water, but each year a few do manage to wreak havoc along the coasts of the Philippines and Japan.

The Atlantic Ocean typically spawns around 12 tropical storms per year, about half of which grow into hurricanes (1933 was the record year, with 21 tropical storms, but 1995 was a close second). The graph in Figure 8.2 shows how tropical storms and hurricanes near U.S. coastlines have historically distributed themselves over the months of the year; although these great storms may arise at any time, they are most common by far in August, September, and October. The islands of the Caribbean and the eastern coast of Central America are particularly vulnerable; only slightly less so are the coasts and barrier islands of the Gulf of Mexico and the eastern states from Florida to the Carolinas. But even as far north as Rhode Island the coastlines are not immune; on September 21, 1938, for instance, a fast-moving hurricane swung to the north and claimed 2,000 lives in unprepared New England. A list of notable East Coast tropical storms and hurricanes is included in Appendix C.

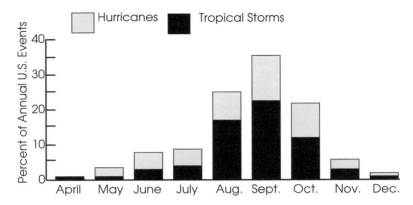

Figure 8.2. Relative frequencies of U.S. tropical storms, and those that intensify to hurricanes, by month. (Based on 868 events over the twentieth century.)

Hurricanes arise out of a relatively rare combination of meteorological
conditions. To begin with, they need lots of atmospheric moisture. Not only
must the sea be warm enough to sustain a high evaporation rate, but the
water must be above 27 °C (80 °F) to a depth of at least 60 meters (200 ft),
if it is not to cool off too quickly as soon as cloud formations shadow it from
the sun and convection rolls the deeper water to the surface. Secondly, the
surface winds in the region of high evaporation need to be converging from
nearly opposite directions. This imparts a circular motion to the air, reduces
the atmospheric pressure, and forces the moisture-laden air upward. As the
water vapor in this rising air column begins to condense into clouds, it
releases latent heat that further warms the rising air currents and causes
them to billow upward to even higher altitudes. At every altitude, however,
there are always preexisting winds. If these winds vary significantly with
altitude, they will quickly rip the storm apart before it forms, so another
requirement is that such preexisting winds aloft must be fairly uniform in
direction and intensity. Meanwhile the rising air currents pull in air from
the surrounding regions, and if this surrounding air is too dry it will quickly
dilute the storm. The fourth condition, then, is that all of the air up to about
5,500 meters (18,000 ft) be fairly humid. In fact, if it is particularly humid,
it will provide an additional infusion of energy into the storm as this added
moisture condenses, and the storm will intensify. Finally, the very top of the
forming storm must have a higher atmospheric pressure than surrounding
regions at the same altitude. If its pressure is lower rather than higher, surrounding air masses will move in and quickly snuff out the storm from the
top.

I suspect that if Mother Nature were to allow a meteorologist to sit in her control booth for a few days, that scientist would have a very difficult time creating a successful hurricane. There are a bewildering number of factors that must combine to make a hurricane develop, and there are an infinity of ways something can go wrong. In fact, even with Mother Nature herself at the controls, fewer than 10% of tropical disturbances ever develop into hurricanes. When a hurricane does form, however, the phenomenon is usually self-sustaining for a fairly long period: about ten days, on average, but sometimes as long as two weeks. During this lifetime, it is not unusual for the storm to travel a distance of 2,500 to 5,000 kilometers (1,500–3,000 mi).

The diagram in Figure 8.3 shows the basic geometry of a fully developed hurricane. The overall shape is that of a funnel some 1,000 kilometers in diameter at the top and tapering down to around 300 to 500 kilometers at the surface. Note, however, that this is a very broad and flat funnel, whose height of 10 to 15 kilometers measures only about 1% to 2% of its diameter – a proportion that is very difficult to show in true scale on a diagram. In the center is the "eye," a cylindrical region about 30 kilometers across that extends to the top of the storm. If you were standing in the eye, you might see blue sky above, and the air would feel unusually calm (although in fact it is actually dropping down on you from above at a speed of some 10 meters per

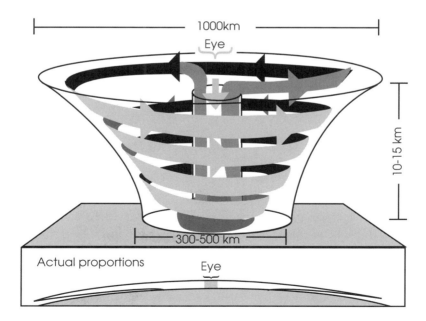

Figure 8.3. The geometry of a fully developed hurricane.

minute). In the wall of the eye, however, there are violent updrafts, and just outside this region lie the most severe winds of the storm.

Barometric pressure is an important aspect of tropical storms, in that it allows us to anticipate the wind speed and the height of the storm swell, or surge. At sea level, barometric pressure normally measures 1013.25 millibars (29.921 in. on a mercury barometer). Tropical cyclones, however, are associated with pressures lower than normal, and the farther the barometer drops, the greater the force that drives the winds toward the central region of lowest pressure. The lowest sea level barometric pressure recorded in the United States was 892.3 millibars (26.35 in. of mercury) during the devastating Category 5 Labor Day Hurricane of September 2, 1935. As I mentioned earlier, the high winds and storm surge claimed 408 lives in that event. The all-time record low was associated with a hurricane that did not strike the United States (although it did spin off twenty-nine tornadoes in Texas, which destroyed some $500 million in property in that state). On September 13, 1988, Hurricane Gilbert roared through the Caribbean, killing 318 in Jamaica and continued on into the Yucatan region of Mexico, altogether destroying $5 billion in property. Wind speeds in that terrible storm peaked at 298 kilometers per hour (185 mi/h) as the barometer plunged to 887.9 millibars (26.22 in. of mercury).

Although winds travel from high pressure regions toward lower pressure regions, they do not follow straight-line paths; instead, due to the Coriolis effect (which is a result of the earth's rotation), they are forced to spiral inward toward the eye of the storm. If this influx of surrounding air should neutralize the central low pressure region, the storm dissipates; this happens fairly quickly when the vortex is over land. Over tropical waters, however, the high winds continually pick up additional moisture, which sustains the process that created the central low pressure region in the first place. A hurricane over water may be viewed as a giant turbine engine whose energy source is solar radiation and whose working substance is warm, humid air. Once started, this engine continues to roar for as long as there is a continued supply of heat and humidity.

Hurricanes begin their lives traveling from east to west, because this is the direction of the prevailing winds in the tropics. They then veer off toward higher latitudes, in many cases changing their direction toward the east before dissipating over the colder, more northerly waters or over land. The map in Figure 8.4 shows the paths of some of the major East Coast hurricanes of the twentieth century.

A hurricane's rate of movement along its path, usually referred to as its *storm speed*, has no direct relation to the wind speeds developed within the

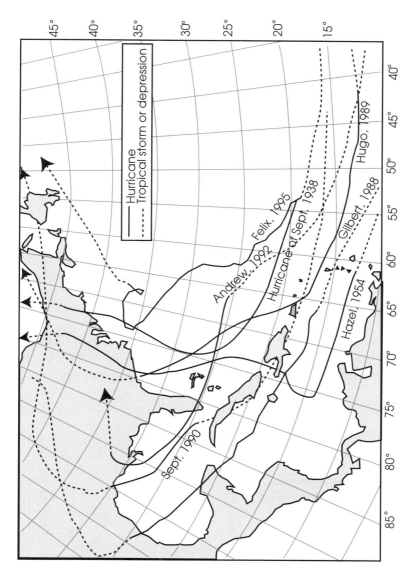

Figure 8.4. Paths taken by some representative hurricanes.

storm. In other words, a Category 4 hurricane may move faster or more slowly than a Category 1 hurricane, and both may speed up and slow down erratically. Some hurricanes even remain stationary for a period of time. It's difficult to state a meaningful average, but storm speeds in the neighborhood of 16 kilometers per hour (10 mi/h) are not uncommon, speeds of 40 kilometers per hour (25 mi/h) are not unusual, and hurricanes have been known in rare cases to travel at storm speeds as high as 120 kilometers per hour (75 mi/h). Notice that a storm moving at 80 kilometers per hour (50 mi/h) will travel as far as 1,900 kilometers (1,200 mi) in a twenty-four-hour day. Such fast-moving storms are a particular menace to both shipping and coastal communities, because they provide so little time for people to get out of their paths.

Every hurricane raises a bulge in the waters beneath it, a phenomenon known as a *storm surge* or a *storm swell*. This effect is easily simulated by blowing over a strip of paper as in Figure 8.5; the loose end of the paper will rise into the moving airstream. In a Category 3 hurricane, the sea level may rise as much as 3.6 meters (12 ft), which is more than enough to cause serious flood damage along most shorelines. Worse, it must be remembered that a storm surge rides on top of the existing tides, and that wind-driven waves ride on top of the combination. The sum total can be devastating, as was testified by the terrible Galveston disaster of 1900 (discussed in detail in Chapter 5). Hurricanes with high storm speeds don't have enough time to raise a

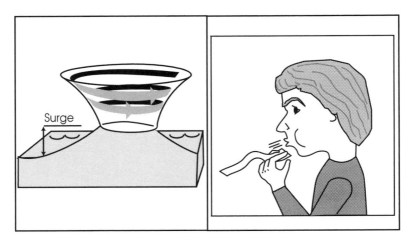

Surge

Figure 8.5. Origin of the storm surge. High winds and low pressures lift the sea into a bulge, just as blowing over a piece of paper lifts it into the airstream.

large storm surge; of all the devastation that Andrew dealt in 1992, for instance, very little was due to flooding from the sea.[5] In this respect, Andrew stands in contrast to the Galveston disaster, where that (unnamed) hurricane had lower sustained wind speeds but progressed forward so slowly that it had time enough to raise a huge storm surge that swallowed up the entire island for several terrible hours.

Although slow-moving hurricanes present the greater threat from storm surges, fast-moving hurricanes present the greater threat from wind damage at points to the right of the path of the eye. This happens because the forward motion of the storm as a whole combines with the counterclockwise motion of the winds within it. At points to the right of the path of a hurricane's eye, these two motions are in the same direction and they add; at points to the left of the eye the wind speed is opposite to the storm speed, and they subtract. Suppose, for instance, that a hurricane has a sustained counterclockwise wind of 160 kilometers per hour (100 mi/h) and is traveling due west with a storm speed of 40 kilometers per hour (25 mi/h). Then structures just north of the eye of the storm will experience winds of 200 kilometers per hour (125 mi/h), while just south of the eye the buildings are being hit with winds of only 120 kilometers per hour (75 mi/h). In this case, the highest storm surge will also occur north of the path of the eye. This little bit of vector addition carries serious implications for evacuation planning, for one thing that should certainly be avoided is having people move from regions at the left of the eye to regions at the right of the eye. Yet even though the principle is simple enough arithmetically, it becomes bewildering in its application. Hurricanes often change direction without notice, we're not very good at predicting just where a hurricane's eye will make landfall, and our highways don't always run in the directions that would be optimal for a specific evacuation. Usually the only realistic evacuation strategy is just to get everyone inland from the shorelines, where no one should ever attempt to ride out a hurricane.

Meanwhile, within every hurricane are bands of thunderstorms, which dump several inches of rain per hour while kicking up a great deal of turbulence near the ground. Moreover, because the speed of the wind increases as it spirals in toward the wall of the eye, pockets of turbulent air are occasionally caught between a pair of wind streams that are traveling at different speeds. The effect is the same as rolling a pencil between your palms: a rapidly twisting vortex of air. These short-lived phenomena are referred to as *miniswirls* and *microbursts*. They are typically only 15 to 60 meters (50–200 ft) in diameter, but for a few seconds they can develop wind speeds of more than 300 kilometers per hour (200 mi/h). Anything in the path of a miniswirl is instantly destroyed. After-the-fact evidence suggests that as many as one hundred of these transient vortices of destruction were generated when Andrew struck southern Florida. To date, however, there are no confirmed reports of anyone witnessing one, and for good reason: The fury of a full-blown hurricane is just not an environment conducive to making objective observations.

My purpose in this entire discussion has been to convey some sense of the tremendous complexity we encounter in trying to understand what goes on in a hurricane, before anyone even thinks about writing mathematical equations and computer programs to describe the phenomenon. Hurricanes are fraught with uncertainty, some rooted in our own ignorance, but perhaps some intrinsic to the phenomenon itself. And, of course, that which is inherently uncertain cannot be predicted.

The Life and Death of Hurricane Emily, 1993

Natural disasters had a banner year in the United States in 1993. On March 13 and 14, a major blizzard struck the Northeast, claimed 200 lives, and damaged hundreds of thousands of buildings (most of which collapsed under the weight of a meter or more of snow). Then, beginning in late June and extending through early August, the great flood of 1993 inundated more than 8 million acres of farmland in the basins of the Missouri and Mississippi rivers, killed 50 people, left 70,000 homeless, and caused $12 billion in property damage. Coming on the heels of these two major disasters, plus a bombing at the World Trade Center in New York, Hurricane Emily did not make big news; her eye never did make landfall, and she claimed just 2 lives and destroyed only about $10 million in property. Emily does, however, provide us with an representative case study of how hurricanes develop, travel, and decline.

Table 8.3 lists the National Weather Service data for Emily, at intervals of roughly six hours, for the fifteen-day period from August 22 to September 6, 1993. Although Emily spent most of her life as a tropical depression, she flared into a hurricane for the seven days from August 27 until September 3, peaking out at Category 3 for about thirty-six hours. In examining Table 8.3, we can make several observations:

1. Although the wind speed was fairly constant at 35 mi/h for the first ninety hours of this data record, the storm speed varied from 0 to 14 mi/h in this same period. Storm speeds are often erratic, and they change independently of the wind speed.

Table 8.3 *Tracking data from Hurricane Emily, 1993 (National Weather Service, Miami.)*

NAME: EMILY
DATE: 9/5/1993 23:00 edt

UPDATES TO SHOW CURRENT POSITION OF EMILY ARE AVAILABLE. FORECASTS
HAVE ENDED DUE TO EMILY'S DISTANT POSITION FROM THE U.S. IN THE
NORTH ATLANTIC OCEAN.

SP STORM POSITION DATA

MM/DD/YY	HH:MM ExT	LAT	LONG	MPH MVMNT	MPH WND	MB GST	PRESS	COMMENTS
08/22/93	17:00 EDT	20.0	53.3	W 12	35		1008	585 MI NORTH LEEWARD IS
08/22/93	23:00 EDT	20.3	53.5	W 7	35		1008	1085 MI EAST OF SAN JUAN
08/23/93	05:00 EDT	21.7	55.4	WNW12	35		1008	930 MI SE OF BERMUDA
08/23/93	11:00 EDT	23.0	57.5	NW 14	35		1015	780 MI S-SE OF BERMUDA
08/23/93	17:00 EDT	23.7	58.0	NW 14	35		1015	725 MI SSE OF BERMUDA
08/23/93	23:00 EDT	24.7	58.3	NW 13	35		1015	650 MI SE OF BERMUDA
08/24/93	05:00 EDT	25.4	59.3	NW 13	35		1015	580 MI SSE OF BERMUDA
08/24/93	11:00 EDT	27.3	59.8	NW 8	35		1016	450 MI SE OF BERMUDA
08/24/93	17:00 EDT	28.3	60.2	NNW 8	35		1016	390 MI SE OF BERMUDA
08/24/93	23:00 EDT	28.4	60.1	N 1	35		1014	385 MI SE OF BERMUDA
08/25/93	05:00 EDT	28.6	60.6	NW 2	35		1014	350 MI SE OF BERMUDA
08/25/93	11:00 EDT	28.1	60.6	STNRY	35		1016	380 MI SE OF BERMUDA
08/25/93	17:00 EDT	28.6	60.1	N 4	35		1016	380 MI SE OF BERMUDA
08/25/93	23:00 EDT	28.5	60.8	STNRY	35		1014	355 MI SE OF BERMUDA
08/26/93	05:00 EDT	28.6	61.2	W 3	35		1014	330 MI SE OF BERMUDA
08/26/93	08:00 EDT	27.2	61.0	W 2	70		1008	420 MI SE OF BERMUDA
08/26/93	11:00 EDT	27.2	61.4	W 3	75		1005	415 MI SE OF BERMUDA
08/26/93	17:00 EDT	27.4	62.5	W 3	75		1005	365 MI SSE OF BERMUDA
08/26/93	23:00 EDT	26.6	62.9	W 7	70		1008	410 MI SSE OF BERMUDA
08/27/93	05:00 EDT	26.6	63.9	W 8	70		1008	400 MI S OF BERMUDA
08/27/93	11:00 EDT	26.5	64.0	W 7	70		993	405 MI S OF BERMUDA
08/27/93	17:00 EDT	26.5	64.9	W 9	70		990	405 MI S OF BERMUDA
08/27/93	23:00 EDT	26.6	65.8	W 10	80		981	400 MI S OF BERMUDA
08/28/93	05:00 EDT	26.9	66.6	WNW 9	80		982	850 MI E OF WPALM BEACH
08/28/93	11:00 EDT	27.6	67.5	WNW 9	80		981	800 MI E OF WPALM BEACH
08/28/93	17:00 EDT	28.3	67.9	NW 9	85		975	785 MI ESE OF S. CAR COAST
08/28/93	23:00 EDT	29.0	68.7	NW 10	85		975	690 MI ESE OF S. CAR COAST
08/29/93	05:00 EDT	29.8	68.9	NW 10	80		978	630 MI ESE WILMINGTON, N.C.
08/29/93	11:00 EDT	30.5	69.5	NW 9	80		978	480 MI SE CAPE HATTERAS N.C.
08/29/93	17:00 EDT	31.2	70.1	NW 9	80		979	420 MI SE CAPE HATTERAS N.C.
08/29/93	23:00 EDT	31.6	70.5	NW 9	85		978	390 MI SE CAPE HATTERAS N.C.
08/30/93	05:00 EDT	31.8	71.1	WNW 8	85		977	350 MI SE CAPE HATTERAS N.C.
08/30/93	11:00 EDT	31.9	71.8	WNW 7	85		975	325 MI SE CAPE HATTERAS N.C.
08/30/93	14:00 EDT	31.9	72.2	W 7	85		973	285 MI SE CAPE HATTERAS N.C.
08/30/93	20:00 EDT	32.2	73.0	WNW 8	95		971	260 MI SE CAPE HATTERAS N.C.
08/30/93	23:00 EDT	32.5	73.5	WNW 9	100	120	972	225 MI SE CAPE HATTERAS N.C.
08/31/93	02:00 EDT	32.8	74.0	WNW 9	100	120	971	190 MI S-SE CAPE HATTERS
08/31/93	05:00 EDT	33.2	74.5	NW 9	100	120	970	155 MI S-SE CAPE HATTERAS
08/31/93	08:00 EDT	33.5	74.7	NW 9	105	125	965	130 MI S-SE CAPE HATTERAS
08/31/93	11:00 EDT	34.1	74.8	NNW 9	105	125	965	90 MI SSE CAPE HATTERAS
08/31/93	13:00 EDT	34.4	75.1	NNW 9	105	125	967	70 MI SSE CAPE HATTERAS
08/31/93	14:00 EDT	34.6	75.2	NNW12	105	125	967	50 MI SSE CAPE HATTERAS
08/31/93	17:00 EDT	35.2	75.1	N 13	115	140	964	25 MI E CAPE HATTERAS
08/31/93	21:00 EDT	35.7	75.0	N 12	115	140	960	95 MI SSE VIRGINIA BEACH, VA
08/31/93	23:00 EDT	36.0	74.7	NNE11	115	140	960	90 MI SE VIRGINIA BEACH, VA
09/01/93	02:00 EDT	36.6	74.4	NNE13	115	140	962	95 MI E VIRGINIA BEACH, VA
09/01/93	05:00 EDT	37.1	73.9	NE 13	115	140	963	125 MI EAST VA BEACH, VA
09/01/93	11:00 EDT	38.0	71.9	ENE17	115	125	967	250 MI S-SW NANTUCKET
09/01/93	17:00 EDT	38.5	69.8	ENE18	110	125	968	190 MI S NANTUCKET
09/01/93	23:00 EDT	39.4	67.3	ENE19	110	115	968	310 MI S YARMOUTH, NOVA SC.
09/02/93	05:00 EDT	39.1	64.5	E 23	105		970	335 MI SSE YARMOUTH, NOVA SC.
09/02/93	11:00 EDT	39.4	61.9	E 20	90	110	970	
09/02/93	17:00 EDT	39.1	60.0	E 21	90	110	970	625 MI SSW CAPE RACE NFNDLND
09/02/93	23:00 EDT	38.7	58.3	E 20	85	105	975	610 MI S-SW CAPE RACE
09/03/93	05:00 EDT	37.7	56.7	ESE17	80	110	977	400 MI SSE SABLE ISLAND
09/03/93	11:00 EDT	37.3	58.0	ESE 0	75	90	980	690 MI SSW CAPE RACE NFL
09/03/93	17:00 EDT	37.0	57.6	SE 3	65	80	987	710 MI SSW CAPE RACE NFL
09/04/93	05:00 EDT	35.2	57.2	S 6	45	55	1002	825 MI SSW CAPE RACE NFL
09/04/93	11:00 EDT	35.5	57.5	NE 0	40	50	1002	800 MI SSW CAPE RACE NFL
09/04/93	17:00 EDT	36.0	57.0	NE 5	35	45	1005	760 MI SSW CAPE RACE NFL
09/04/93	23:00 EDT	37.3	56.8	NE 7	35	40	1004	675 MI SSW CAPE RACE NFL
09/05/93	05:00 EDT	37.7	56.1	NE 7	30		1006	635 MI SSW CAPE RACE NFL
09/05/93	11:00 EDT	38.2	55.3	NE 8	30		1006	590 MI S CAPE RACE NFL
09/05/93	17:00 EDT	39.1	53.8	NE 10	30		1008	525 MI S CAPE RACE NFL
09/05/93	23:00 EDT	38.9	52.5	ENE10	30		1008	540 MI S CAPE RACE NFL
09/06/93	05:00 EDT	39.0	50.0	ENE12	30		1010	550 MI SSE CAPE RACE
09/06/93	11:00 EDT	39.5	48.4	ENE15	30		1014	580 MI SSE CAPE RACE

Source:
National Weather Service Miami
The last information release for Hurricane Emily

2. As the barometric pressure dropped, the wind speed increased. The lowest barometric pressure, 960 millibars, corresponds to the highest wind speed of 115 mi/h, and to the highest gusts of 140 mi/h. Once again, this is generally the case: The highest winds occur when the barometric pressure is lowest.

3. The storm began by traveling to the west, then gradually veered to the north, then toward the east. Hurricanes always curve to the north, and most end their lives traveling in a northeasterly direction.

4. Emily's highest storm speed was 23 mi/h, while she still had dangerous wind speeds but was fortunately traveling away from coastal regions. In one twenty-four hour period (beginning at 11 PM on September 1), Emily covered a distance of about 480 miles.

5. Emily did not break into the hurricane category until she was at a latitude of 27.2 °N. This is even farther north than Andrew's emergence as a hurricane the previous year.

6. Emily lost intensity over the cooler waters near a latitude of 40.0 °N.

7. No variation in intensity is apparent between day and night. The seas have too large a thermal capacity to heat or cool quickly when the sun rises and sets.

Figure 8.6 shows an enhanced satellite photograph of this hurricane at about 8 AM on August 31, 1992. At this time, Emily was roughly 130 miles south-southeast of Cape Hatteras, moving northwestward at about 9 miles per hour, and still intensifying. The indications were very clear that a Category 2 or 3 hurricane would strike land within fourteen hours, somewhere north of Wilmington, North Carolina. Most of the barrier islands of the Outer Banks would be to the right of the eye and would therefore be bashed with the highest winds and the greatest storm surge. Because memories of Andrew were still fresh in the public mind from the previous year, almost everyone along the threatened coastlines and islands heeded the evacuation warnings.

In the next six hours, the storm abruptly shifted its direction to the north-northwest, and the storm speed increased to 12 miles per hour. Emily's eye was now just 50 miles out to sea and heading directly toward Cape Hatteras. In the neighborhood of the historic lighthouse, rain fell in opaque sheets, and huge breakers crashed on the beaches. Anyone who had not already evacuated had no choice but to ride it out. The accelerated storm speed, however, also increased the Coriolis force that slowly curves most hurricanes' paths to the right. Three hours later, at about 5 PM on August 31, the

Figure 8.6. Satellite image of Hurricane Emily, 1993. (Photo Courtesy Burt Baker, SeaSpace Corp.)

eye was still 25 miles offshore from the cape but now was headed due north. The fact that the hurricane had just broken into Category 3 with winds of 115 mi/h was of no grave consequence; the Outer Banks now lay on the back side of the vortex, and the situation there would get no worse. Most of the storm surge was well out to sea, slightly to the right of the storm's path. Hatteras experienced battering winds of 75 miles per hour, a surge of more than 5 feet, beach erosion, and wind-damaged houses, but nothing compared to what might have been.

Had Emily continued to head due north at this point, she still would have dealt considerable destruction along the coasts of New Jersey, Long Island, and perhaps Rhode Island and eastern Massachusetts. Instead, she fortunately continued to curve to the east and reached her greatest intensity when she was safely out to sea. Once she left the Gulf Stream, the waters were not warm enough to sustain her. Her winds gradually slowed, her barometric pressure began to climb back toward normal, and by midday on September 3, Emily was downgraded to a tropical storm. In this mode she made a series of curious twists and turns, none of which did anything to help her recover her strength, and on September 4 her intensity was downgraded again to a tropical disturbance. Eventually she died alone far out to sea, long after everyone had ceased to pay attention.

Winds and Structures

During a winter storm in December of 1879, a passenger train rumbled onto a tall iron trestle bridge spanning the Firth of Tay in Scotland. The bridge itself, with its open trusswork, allowed the howling winds to pass through with little obstruction. The train, however, acted like a very broad sail that caught the wind, transmitted this horizontal wind load to the anchored rails, and toppled both the bridge and the train into the angry waters below. At least eighty travelers perished in that bridge failure.

The engineering community does tend to learn from its mistakes (many engineering students even take a course in "Failures"), and today no railroad bridge would ever be designed without accounting for the increased effect of the wind while a train was crossing. Still, the field of wind engineering is a very difficult one where subtle mistakes do occur, sometimes through human blunder, but more often through shortcomings in the available data or the accepted mathematical models. We still don't know all there is to know about how nature's winds interact with man-made structures.[6]

Most of the fatalities and economic losses from hurricanes, however, result not from failures of engineered structures but rather from failures of conventional dwellings. People don't normally hire an engineer to design their new home (although in some high-risk areas an engineer's stamp of approval may be required). Instead, most builders just follow the local building codes, which in turn reflect the materials and construction standards that have a historical track record of being effective. Given that codes are reductionistic in nature, this approach is bound to lead to occasional unpleasant surprises. It is possible, for instance, for two identical houses at different elevations or with different compass orientations to fare quite differently in the same windstorm. It is also possible for subtle differences in roof pitch or window placement to play a major role in whether a house survives an onslaught of high winds.

In even the very worst storms, and even out in an open field, the wind velocity right at ground level is zero. What is referred to as the "surface wind speed" is not really the wind at the surface of the earth but rather at a standard height of 10 meters (33 ft) above the ground, where the anemometers are usually mounted. From this point, the wind speeds generally increase with altitude; thus, the winds measured by aircraft or balloons are not the same as the winds encountered by the structures below them. The wind speed referred to in the Saffir-Simpson Scale of hurricane intensities (Table 8.2) is the "sustained speed" at a 10-meter height. The U.S. National Weather Service uses the average speed over a 1-minute period to determine this sustained speed, but gusts (as we have seen with Emily in Table 8.3) can range considerably higher.

High winds generate several types of forces on a structure. The most obvious is the inertial force, which arises when a face of the structure stops (or alters) the forward momentum of the moving air mass. This is the force you feel on your outstretched palm if you stick your hand out the window of a moving car. Clearly, this inertial force depends on the orientation of your palm; it is greatest when you hold the palm perpendicular to the airstream – the orientation where it presents the greatest frontal area. This inertial force also increases with the wind speed. Unfortunately, there are other factors that also come into play and make accurate predictions difficult. As a representative value, however, one can figure that an object presenting a frontal area of 1 square foot will experience roughly 1 pound of inertial force in a 20 mile per hour wind.

Does this mean that we would get 2 pounds of force in a 40 mile per hour wind, 3 pounds at 60 miles per hour, and so on? No, it's not this simple (and if this indeed were the case, buildings would very seldom suffer serious wind damage). What actually happens is that every doubling of the wind speed quadruples the inertial force, and tripling the wind speed increases the inertial force by a factor of 3^2, or 9. This relationship can be summarized as follows:

$$\text{Wind inertial force } \textit{is proportional to } (\text{windspeed}).^2$$

Thus, for a house whose face presents an area of 400 square feet, we can predict the following approximate lateral inertial forces as the wind speed increases:

Wind speed (mi/h)	Inertial force (lb)
20	400
40	1,600
60	3,600
80	6,400
100	10,000
120	14,400
140	19,600
160	25,600

Notice that a building that survives an 80 mile per hour wind would need to be *more than twice as strong* to survive a 120 mile per hour hurricane.

What I have just described is a *steady-state inertial effect:* the force of the wind blowing more or less continuously against the outside of a building. It should not be confused with the *blast effect,* where a window or door suddenly breaks open and the wind exploding into the structure destroys it from the inside. No realistic amount of structural engineering can safeguard a building against the blast effect. It makes much more sense to take every possible precaution to ensure that windows and doors will not break or fly open during high winds.

In addition to its inertial force, a high wind will also develop aerodynamic forces on various surfaces of a building. These forces and their interactive effects, though simple to understand, are particularly difficult to predict. I've already mentioned how high winds raise a bulge (a storm surge) on the ocean's surface. In like manner, winds will generate an upward force ("lift") on any roof that is pitched at an angle less than about 25 degrees, and an outward force on any wall parallel to the wind stream. In addition, a force of turbulent drag (sometimes called "suction," although I personally dislike this term) will develop on any surface facing downwind, including the downwind slope of a steeply pitched gable. Some of these effects are shown in the diagrams in Figure 8.7. The pagoda-style roof is an interesting case, in that it can develop a negative lift at the overhanging portions of the roof. This actually helps hold such a roof in place during a high wind, as is testified by the large number of centuries-old Japanese and Chinese structures that have survived numerous typhoons in their lifetimes.

In any major windstorm, then, some parts of a building will be pushed inward while other parts are being pulled outward. This combination of pushing in one place while pulling in another is particularly abusive and leads directly to many structural failures. Moreover, as wind direction shifts

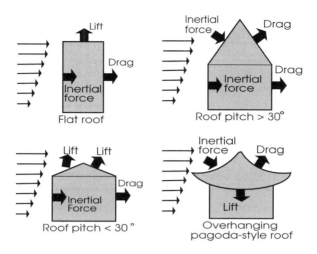

Figure 8.7. The forces transmitted to buildings by high winds depend on the shape and orientation of the specific structure.

(and it usually does during any major storm), many of the forces reverse direction. Often a building will be weakened during the first half of a hurricane, then will collapse when the winds reverse their direction after the eye passes through.

What can a do-it-yourself homebuilder do? Several things. Don't depend on gravity to hold the house in place; instead, anchor everything to the foundation. Use tension straps at all major joints, particularly between the roof rafters and the walls. Make sure any masonry chimney is reinforced with steel. Use plywood rather than chipboard for roof and wall sheathing. Install working shutters on windows. Have exterior doors open outward rather than inward. And even then, when a hurricane warning is issued, evacuate to a reinforced shelter that has been specifically engineered to withstand high winds. If your house has been built right, it will probably still be standing when you return. But if it has structural deficiency, you don't want to find out about it in the midst of a hurricane.

Tornadoes

At high speeds, most liquids and gases tend to swirl into eddies and vortices, rather than flowing smoothly. Usually these regions of circular motion break up soon after they form, and the overall motion becomes turbulent and unstructured. Occasionally, however, a vortex achieves a kind of metastability that permits it to survive for many minutes or even an hour. The most dramatic and devastating meteorological example of this phenomenon is the tornado.

It is difficult to describe a "typical" tornado, other than to say that it is a rapidly whirling funnel cloud that emerges from a violent thunderstorm and reaches to the ground. Some touch ground for only seconds, others for up to an hour, and some hopscotch: a few minutes here, a few seconds there. Some are only a few meters across, others are several hundred meters in diameter even at their base. Their colors vary: Some are light gray, others dark, and when one of them crosses a newly plowed field it quickly assumes the brown color of the earth it scoops up. Some form over water rather than land, and these are called "waterspouts." Occasionally tornadoes are observed in pairs that conduct an eery, twisting dance around each other. They also occur in families: a group of small tornadoes orbiting a large central tornado. Unlike hurricanes, tornadoes can twist either clockwise or counterclockwise (although the latter are apparently more frequent). What they do have in common is that they are all extremely destructive. It is quite impractical to design a house that will survive a direct strike (or even a near-miss) by a tornado.

The United States has a virtual monopoly on tornadoes; a few do spill over into south-central Canada and northeastern Mexico, but only rarely do they occur anywhere else in the world. Of the contiguous forty-eight states, none is immune. The highest frequencies are in the midwestern and southeastern states, during the months of March, April, and May. Appendix D lists the major "killer tornadoes" of the last one hundred years. Contrary to the customary notion that tornadoes are a threat only in Oklahoma, Kansas, and Texas, a quick glance at the list shows that they have claimed many hundreds of lives and have destroyed millions of dollars in property in numerous states. The photograph in Figure 8.8 shows a fairly typical Texas tornado, with a cloud of debris rising from its base.

Tornado intensities are rated on the Fujita Scale (Table 8.4), according to their wind speeds. Although this scale was originally designed to range from F-0 to F-12, it is currently believed that no tornado ever exceeds an F-5 intensity (a peak wind speed of 318 mi/h). Even at the lower F numbers, though, these are still extremely high winds. Tornado damage is caused not only by the direct effects of these winds, but also by several collateral effects: (1) the intense updrafts within the funnel (estimated at up to 200 mi/h), (2)

Figure 8.8. A tornado approaching Dawn, Texas, on May 9, 1982. (Photo Courtesy Tim Marshall.)

the funnel's extremely low atmospheric pressure, which may be one or two pounds per square inch less than the barometric pressure of the surrounding atmosphere, and (3) the effects of flying debris.

The photographs in Figure 8.9 shows two views of the aftermath of a compact F-3 tornado that roared through a newly constructed residential development in Limerick, Pennsylvania, in June of 1994. About a dozen houses were seriously damaged or destroyed, while many adjacent houses suffered no damage at all. It's apparent from the photographs that the devastated structures exploded outward, not inward: Complete roofs are gone from several of the houses, some exterior walls are completely missing, and considerably more damage is apparent at the upper levels than close to the ground. Multipane windows seem to have fared pretty well unless they were struck directly by flying debris. Although all of this is consistent with what I've already described as the expected effects of high winds on structures, one additional point merits specific attention. Tornadoes are so highly localized that the barometric pressure plummets dramatically in a matter of seconds when one roars past. If doors and windows are tightly shut (and they probably will be, given that it's raining cats and dogs), the air pressure

Table 8.4. *The Fujita Scale for tornado intensities*

Category	Damage	Wind speed	Typical effects
F-1	Light	40-72 mi/h	Damage to trees, free-standing signs, some chimneys.
F-2	Moderate	73-112 mi/h	Roofs damaged, mobile homes pushed from foundations, moving cars swept off roads.
F-3	Severe	113-157 mi/h	Large trees uprooted, roofs torn from houses, mobile homes demolished, boxcars toppled, collateral damage from flying debris.
F-4	Devastating	158–206 mi/h	Well-constructed houses explode, cars become airborne, collateral damage from large missiles.
F-5	Incredible	261-318 mi/h	Buildings lifted from foundations and disintegrate in the air, cars carried distances greater than 100 m.
F-6 and up		Over 318 mi/h	Currently believed nonexistent.

Note: About 75% of all tornadoes exceed F-1 intensity; 35% exceed F-2; 9% exceed F-3; 2% exceed F-4; and fewer than 2% reach F-5.

inside the house does not have a chance to equalize with the external drop in atmospheric pressure. As a result, each square inch of external wall experiences an outward force of as much as a pound or two. Considering that an 8-foot by 30-foot wall has about 35,000 square inches of surface, a near-miss by a tornado can easily generate an outward force of 18 to 35 tons on a single wall. It's no wonder that a house will explode under these conditions. In fact, you can buy or rent tornado videotapes that, if examined closely, will actually show buildings exploding as a result of this rapid pressure imbalance.

Figure 8.9. Aftermath of an F-3 tornado in Pennsylvania, June 21, 1994. (Photos Courtesy Limerick Fire Company, Limerick, Pa.)

A building is much less likely to be destroyed by a tornado if a few doors or windows are left open on the downwind side. Although this won't protect against a direct strike, it does allow the pressure to equalize if the twister happens to pass very close. Unfortunately this is not always practical advice, because the direction of the wind may not be apparent at the time the occupants of the home evacuate or move to the basement. If one should miscalculate and leave upwind-facing windows open, the blast effect can be every bit as destructive as the explosive effect of captured inside air. Most experts recommend shutting the windows and letting the odds fall where they may.

Human survival during a tornado depends on not being struck by flying debris and on not being physically swept into the funnel cloud; in other words, on avoiding the high winds themselves. In a house, occupants should always get into the basement or a first-floor room with no exterior walls. Anyone in a car should get out, crouch or lie in a low place (where the wind speed is always near zero), and not complain about getting soaked in the downpour. Because the conditions that create tornadoes also generate a great deal of lightning, it's important to avoid trees and to discard metal objects like umbrellas or golf clubs. Even in the open, there is no guarantee that a victim will see the tornado coming; often the surrounding rain and hail are so intense that the tornado structure itself is totally obscured. Survivors, however, never fail to mention that they *heard* the twister; they most often compare the sound to that of a speeding freight train bearing down on them.

But there is yet another hazard. In my part of the country, north-central Pennsylvania, visitors are often surprised at the large number of traffic accidents that involve collisions with deer. Don't our rural Pennsylvania drivers pay attention to the road? As a matter of fact, the typical scenario runs like this: The driver sees a deer, brakes or swerves to avoid it, watches the animal disappear into the woods, then collides with a second deer that's been following the first. Tornadoes, like deer, often travel in groups, and successfully avoiding the first is no guarantee of one's fate with the second (or third, or fourth) of the party. The turbulent conditions that give rise to tornadoes usually extend over a broad region, and under severe conditions a long line of thunderclouds can spawn dozens of twisters. Although the number of tornadoes (even if it's just one) cannot be predicted at present, it is easily determined after the fact. We do not need to see a tornado in action to know that one's passed through: Cars will be flipped and/or carried great distances, roofs will be missing from houses, and large trees will be twisted off at midtrunk. Determining the number of separate tornadoes in an event requires no more than carefully assimilating all of the damage records after the fact. We know, therefore, that the most serious historical tornado disasters have resulted from multiple twisters. On February 19, 1884, for instance, around 800 people in seven states died from the effects of about 60 tornadoes, and on April 3 and 4, 1974, 350 deaths in five states resulted from 144 tornadoes. We must consider the possibility that some of these victims thought the danger was over when it wasn't. In fact, any time a single tornado is seen or reported, it makes sense to assume the presence of additional tornadoes in the region. Don't assume it's over until the skies turn blue.

The Limitations of Forecasting

No, I'm not going to delve into highs and lows and isobars and squiggly lines on weather maps; we've all seen these enough times on television, and we know that regardless of how scientific we try to make them, forecasts often remain frustratingly inaccurate. What I do want to talk about is why we can't achieve 100% accuracy in our weather predictions, and why it may turn out that we never will.

All forecasting is based on "extrapolation," which is the mathematical term for extending a series of data points beyond the last known value. The most naive approach to predicting the future of anything is to make a linear extrapolation. Put a pot of water on the stove, for instance, and measure its temperature increase during the first minute, then the second. Suppose we find that the temperature increases from 25 °C to 30 °C to 35 °C. On this basis, we might predict that the temperature will continue to rise by 5 °C each minute, so that after an hour the pot of water should be at 25 °C + (60 × 5 °C), or a total of 145 °C. This is good mathematics, but it is naive science. Why? Because once the water reaches 100 °C, it begins to boil and it can get no hotter. After an hour of heating, it's still at 100 °C, assuming any is still left in liquid form.

A scientific prediction requires that we do much more than just extrapolate numbers in linear fashion; instead, we first need to truly understand the underlying physical processes. In the simple case of the pot on the stove, we need to know how the incoming energy is distributed between the water and its container (and how this distribution depends on temperature); we need to know how both the water and the pot transmit heat energy to the surrounding air (a rate that may also be temperature-dependent); and we need to know what happens when the water changes phase from liquid to gas. If these processes can be described mathematically and we know how to solve the resulting equations, we may then come close to predicting the temperature of the water at any future point in time. A great deal of intellectual work to predict a very simple physical event!

Weather, as we have seen, is considerably more complicated than a pot of water on a stove. Dozens of physical processes are involved in weather, and some (turbulence, for instance) do not yield to simple mathematical description. Weather also covers broad regions of the earth (winter Arctic winds may affect future weather in Virginia, for instance), so that the mathematical models must deal not only with time sequences but with place sequences. The more variables we introduce, the more complete and less naive are our mathematical models. Yet there are drawbacks to increasing the sophistication of our equations: As we increase the number of variables, we simultaneously increase the amount of measurement data we require for the computations, and we increase the computation time.

Computers, of course, have revolutionized our ability to perform complex and tedious calculations, and to do so at lightning speed. At present, however, computers do not actually think. Every computer requires that a human give it a set of appropriate mathematical instructions, and then that a human feed it the relevant data needed to perform the programmed computations. The program and the data are two quite different issues, and they originate in quite different ways. Yet a glitch in either will lead to an erroneous prediction, a forecast of a natural event that fails to follow the script. The great Albert Einstein recognized this programming problem well before the days of electronic computers when he said, "Insofar as mathematics applies to reality, it is not certain. Insofar as mathematics is certain, it does not apply to reality." The point is that our mathematical models are only that: models. Mathematics is a human invention, and Mother Nature may have something quite different up her sleeve. It is unlikely that any mathematical system will ever render all future events precisely predictable.

Suppose, however, that some devil's advocate did write a computer program that predicted the future position and speed of every molecule in the earth's atmosphere. Molecules themselves interact in relatively simple ways, and the equations governing their motion are not particularly difficult to solve. Such a computer program might be designed to tell us where every molecule is going, and how fast it's traveling, at any specified instant in the future. Given such information, we could then calculate the macroscopic wind speed, temperature, humidity, and barometric pressure at that same future instant – which is to say we could indeed predict the weather to any arbitrary degree of accuracy. Is this a strategy that might bear fruit someday?

Unfortunately, no. Modeling the earth's atmosphere at the level of molecular interactions is quite impossible, even in theory. Because the earth's atmosphere contains many more molecules than the number of electrons (or photons for that matter) in any conceivable computer, no electronic computation could possibly keep up with the rate at which all of the atmospheric molecules are changing their speeds and positions in real time. There is always a level of detail where the mathematical calculations in any computer grind *more slowly* than the reality being modeled.[7] A program that modeled weather at the level of molecular interactions might succeed in predicting the weather, but it would be weather that happened last week.

It's a fundamental principle, in other words, that no computer can model the behavior of something more complex than its own hardware, unless it models it more slowly than it really happens. Weather forecasts, unfortunately, are useless if the weather itself arrives before the forecast. If we want to avoid this pitfall, we have no choice but to limit the complexity of our models and deal with some entity larger (and fewer in number) than the atmospheric molecules themselves.

Still, we don't really demand extreme precision in our weather forecasts. No one cares, for instance, if a predicted temperature is off by 3 °C, the arrival time of a hurricane is off by a quarter hour, or the height of a predicted storm swell is in error by 20 centimeters or so. Are there still inherent limits on our forecasting ability if we are content to allow our predictions to be a little fuzzy? The answer, of course, is that predictions indeed grow more reliable as we agree to tolerate more "fuzz." This does not mean, however, that anyone has agreed to tolerate sloppy science; what it means is that we are bumping up against some fundamental epistemological issues that limit how good our best science can be. As a result, we can never be guaranteed that the fuzz in our predictions won't sometimes expand to unexpected proportions.

To see how this happens, let's return to our mathematical model of the atmosphere, but instead of modeling the dynamics of individual molecules, let's calculate the average behavior of small groups of molecules. Immediately we deviate from reality, and we get approximate forecasts at best. But this should be okay, because we already agreed that we didn't need extreme precision for a forecast to be useful. So, do we run into any *other* problems if we imagine the atmosphere to be partitioned into a large number of small cubical cells, each described by its average molecular position and speed?

Unfortunately, yes. Suppose, for purposes of computation, that we divide the troposphere into 100 million identical cubical cells. Each of these cells, it turns out, has a volume of about 10 cubic miles, corresponding to a cube some 2.15 miles on a side. These are pretty large from a molecular point of view, each containing around 2.5×10^{37} air molecules, so we've simplified things considerably. In terms of the demands of our mathematical models, we need to physically measure only four thermodynamic coordinates for each cell. (There is some choice of the specific four, but temperature, pressure, density, and chemical composition work pretty well.) Thus, our approximate model ought to predict the weather with reasonable accuracy if we feed it 400 million atmospheric measurements, all of which have been made at the same instant in time. But is this realistic? Far from it. There is simply no way to get such a wealth of simultaneous data by using any foreseeable technology.

Many measurements can be made today using weather satellite instrumentation, global positioning telemetry, ground-based Doppler radar, and standard fixed-station instrumentation, but there is just no way to make several hundred million simultaneous measurements and immediately feed them to a central computer. When the day arrives that we have this level of capability, we will all be more satisfied with most of our daily weather forecasts. But our more dramatic weather anomalies, such as hurricanes and tornadoes, will always be highly sensitive to details that lie obscured within the finite atmospheric cubicles our mathematical models need to work with. When science bumps up against the fundamental limits of what is measurable, and what is describable mathematically, Mother Nature continues to have free reign to deal out future surprises.

Notes

[1] R. Monastersky, Unusual weather spurred Andrew's growth, *Science News*, Sept. 5, 1992, 150.

[2] Most of this account is drawn from articles in the Miami *Sun-Sentinel* and the Boca Raton *News* for the period August 23 through August 30, 1992, plus local news telecasts and personal interviews with several individuals who experienced the event.

[3] S. Borenstein, Mini-swirls, microbursts boosted Andrew's power, Miami *Sun-Sentinel* May 20, 1993, pp. 1A, 12A.

[4] R. Gore, Andrew aftermath, *National Geographic*, Apr. 1993, 2–37.

[5] Andrew's storm surge did reach 16.9 feet at one point in Biscayne Bay, but this extreme was highly localized and may have been partly a resonance effect

related to the shape of the bay. It is extremely unusual for a fast-moving hurricane to generate a significant storm surge.

6   For an overview of this topic, see Henry Liu, *Wind engineering: A handbook for structural engineers* (Englewood Cliffs, NJ: Prentice Hall, 1991).

7   This is not a new idea, and it is not hypothetical. In the late 1960s, I had a job developing mathematical models and computer simulations of manufacturing processes for a major steel manufacturer. Using the slow computers of that time, we frequently found ourselves forced to simplify our mathematical models so the computation time would not exceed the real time for the process being modeled. The limitation is a very real one; even with today's more powerful computers, there is necessarily a level of detail where the complexity of the real world overwhelms the ability of the computer to describe its dynamics in less than real time.

9

Science and Irreproducible Phenomena

The Butterfly Effect

Could a butterfly in a West African rain forest, by flitting to the left of a tree rather than to the right, possibly set into motion a chain of events that escalates into a hurricane striking coastal South Carolina a few weeks later? As bizarre as this premise may sound, research of the last few decades suggests that the answer is yes, and that the effect is hardly limited to butterflies. Many of the large-scale phenomena that threaten human life yet stymie efforts at scientific prediction – tornadoes, earthquakes, volcanic eruptions, and epidemics among them – seem to have one characteristic in common: a sensitive dependence on seemingly innocuous variations in initial conditions. These are phenomena where small disturbances often escalate to larger ones, and if the initial causative agent changes only slightly, the larger effects can differ quite dramatically.

Of course, no one has ever actually observed a butterfly triggering a hurricane. The physical evidence for the butterfly effect is considerably more subtle than this and begins to make sense only through certain theoretical arguments. One fairly compelling line of reasoning runs as follows: If we wish someday to be able to predict future hurricanes, we first need to identify the prerequisite knowledge that would have permitted us to predict today's hurricane. And there ought to be a simple way of doing this: Just run the documentary in reverse. Take lots of data as a hurricane develops, then run it backward in time to see how the event got started in the first place. When we identify how it started, and we know how step C follows from step B and that from step A, we ought to have a program for predicting future events of the same type.

In fact this research strategy has been followed, many times, and it always ends in dismal failure when applied to complex natural phenomena. The problem is that we can see only so much detail, we can measure only to limited precision, and we can compute only to finite accuracy. The swarm of satellites now circling the globe indeed provides us with vast quantities of highly detailed information: in the technical limit, temperature variations of a few thousandths of a degree, and geometrical resolutions approaching 30 centimeters (1 ft) horizontally and a few centimeters vertically. But suppose that a tropical storm develops, and that we play back the data record of the previous few days. What do we find as we go back in time? A smaller storm, and yet a smaller disturbance, then a warm moist windy spot, then a set of atmospheric conditions that looks no different from that at many other locations in the tropics. What is it that whips some of these minor atmospheric fluctuations into full-blown hurricanes, while others disperse after causing no more annoyance than blowing off someone's hat in Africa? It's impossible to say for sure. All we know is that the fundamental agent must be very small, because all of our expensive and sophisticated instrumentation can't detect it. When scientists run up against something very curious like this, they often become whimsical; hence, the "butterfly effect." The premise is that of extremely sensitive dependence on initial conditions. Hummingbirds or flying squirrels would do as well.

The butterfly effect is not limited to a storm's birth but also applies to its future development (which, as we saw in the detailed case of Hurricane Emily, can be quite complicated indeed). Why, for instance, do hurricanes frequently change storm speed and direction? If we could understand this, we might at least improve the accuracy of our predictions of where and when a specific hurricane will make landfall. We now program our computers with dozens of equations to predict the future course of a storm after it is born, and we feed these equations hundreds of thousands of data values that originated in direct readings of sophisticated instruments, yet our predictions just one day into the future are still only marginally accurate at best. What is going wrong?

An insight into the answer has been gained by comparing computer simulations of hypothetical storms whose initial conditions differ in only the tiniest ways. Hypothetical initial data is supplied, and a simulated storm develops on the computer screen, with columns of numbers telling how the wind speed, storm speed, temperature, barometric pressure, and other measurable variables change as time passes and the storm evolves. Here "initial data" does not refer to data from the time of birth of the storm (a time that is, after all, unknown), but to data from some arbitrary instant in time during the storm's development; the data are therefore "initial" only in that they are used to initiate the computation. By itself, of course, such a single computer simulation tells us nothing of any value. But now conduct a second computer run, and a third, then a fourth, with the same computer program but slightly different initial data. How different? Change the temperature profile by a just few millionths of a degree here and there (which is far beyond the limits of physical measurement), or change the wind-speed distribution in the fourth or fifth decimal place (which might mimic a flock of birds alternately flying into and against the wind). Now compare the results of these different computer runs. What do we find? In the first few hours, as one might expect, these simulated storms differ very slightly. But as time passes, it turns out that their behaviors always begin to diverge, and eventually they often develop in quite different ways. One simulated storm may veer northward while another continues westward, one may intensify while another is dying, or one may stand stationary while another gallops toward a shoreline.

Such experiments with computer simulations suggest that the future of a storm is always extremely sensitive to tiny fluctuations in what goes on within it – so sensitive, in fact, that even variations that are too tiny to measure may seriously affect the future course of the event.

Variations of this type need not come from butterflies or birds. Going back three centuries to Isaac Newton's law of action and reaction (One system cannot affect another without the second exerting an effect back on the first), we realize that if a hurricane strikes even a small island, for example, the island itself will affect the future of the hurricane. Newton himself viewed action and reaction as always arising in equal and opposite force pairs, and at any instant in time they do. In light of the butterfly effect, however, a very small reaction by the island at one instant may initiate a larger action-reaction pair at the next instant, and the process may cascade until it ultimately sends a hurricane off on a significantly different future course. Moreover, the butterfly effect suggests that it may take a lot less than a whole island to do this; even a few newly built hotels may have the capacity to affect the future of a particular storm.

Today's scientists have generally accepted this fundamental premise, and within the last decade a number of new scientific journals have appeared that are devoted to the study of "nonlinear dynamics." The term "nonlinear" refers to situations where two agents combine to produce an effect that is quite different from the sum of the two separate effects. As a simple analogy, take two children (boys, in particular), put them in separate rooms full of toys, watch their behavior, then try to use your observations to predict what will happen when you put them together in the *same* room full of toys. Can such a prediction, even in principle, be made with any degree of reliability? Is there any way to anticipate whether the children will cooperate, fight, ignore each other, smash the toys to smithereens, or alternate between these behaviors? At best, it's a tall order. Regardless of your prediction, you wouldn't want to put the little guys together and then walk away in confidence.

Nonlinear natural phenomena – such as storms, earthquake clusters, and epidemics – tend to be wildly unstable. They may seem to be behaving themselves at one instant in time, following at least some statistically predictable pattern; then suddenly they swing into a drastically different mode of behavior, for reasons that are not apparent because the agent responsible for the change was too small to measure or even to notice. Add a little glitch, a metaphorical butterfly, to a complex process, and sometimes you get an outcome no rational person would ever have expected.

In social and political arenas, the counterpart of the butterfly effect has long been recognized and written about. A well-known fable that apparently originated in pre-Elizabethan England runs as follows:

> For the want of a nail, the shoe was lost.
> For the want of a shoe, the horse was lost.
> For the want of a horse, the rider was lost.
> For the want of a rider, the battle was lost.
> For the want of a battle, the kingdom was lost.
> All for the want of a nail.

It's worth noting that the moral to this story is ambiguous. On the one hand, the writer may have been quite pessimistic about our human abilities to plan for anything, given that even the tiniest glitch can lead to the most drastic consequences. On the other hand, the writer may have been advising us to inspect carefully for loose horseshoe nails.

There are many metaphorical butterflies (and loose horseshoe nails) in the world, and at present there is no way of telling which of them may send a large-scale nonlinear system into upheaval. Mother Nature's message, however, remains just as ambiguous as the horseshoe nail fable. Is she telling us that we've bumped up against the limits of what it is possible for humans to know and predict? Or, maybe what she's really telling us is that we need to explore paths to knowledge other than deterministic or statistical numerical prediction. Then again, the real message may be this: Find the right metaphorical butterfly at the right time, snag it and send it on a different course, and perhaps we can occasionally *prevent* a natural disaster.

The Second Law of Thermodynamics

Let me backtrack a bit, to the early 1800s. At that time, Newton's equations had been applied to mechanical systems for well over a century with great success. Several European scientists, most notably Blaise Pascal and Daniel Bernoulli, had given birth to the science of fluid mechanics. Chemistry was emerging as a quantitative science, due largely to the atomic theory proposed by John Dalton around 1803. Many of the newly discovered scientific principles were immediately and successfully applied to technology, and the Industrial Revolution was in full swing.

The steam engines of the day, though technological marvels, had appetites for fuel that would boggle the mind of any modern driver who complains about his or her car's fuel economy. Inventors scrambled furiously to improve their engines' fuel efficiencies, but they struggled without the guidance of a coherent scientific theory that linked mechanics to heat and the chemistry of the fuel. The unanswered question was this: Does nature impose any fundamental limits on what an engine can accomplish, given a specific amount of fuel? If such a limit existed, and if it could be computed, then engineers and inventors would have a yardstick against which to evaluate the performance of any new engine they might design and build.

This question was answered in 1824 by the young French physicist, Sadi Carnot (who, unfortunately, died just eight years later at the age of 36). Carnot established that there is a well-defined limit to the percentage of a flow of heat that can be converted into organized mechanical motion. Ultimately, Carnot's principle came to be called the *second law of thermodynamics*.[1] Soon after its formulation, this principle began to stimulate numerous advances in power engineering,[2] which in turn stimulated additional work by other scientists to develop a deeper understanding of the nature of heat. In particular, great amounts of intellectual energy went into exploring how the behavior of individual molecules of fluids could lead to Carnot's findings.

As a simple but relevant example, think about the following experiment: You have a 1-liter pitcher of cool water at 10 °C and a second 1-liter pitcher of warm water at 70 °C. You pour both pitchers of water into a bucket, and of course the final temperature of the mixture is 40 °C. But now reverse the process, and pour water out of the bucket back into the two original pitchers. Do you get back the 10° water and the 70° water? Of course not. But *why* not? After all, you know that the warm water and the cool water are both in the bucket, because you put them there in the first place!

So let's try to do something a bit more clever to recover the warm and cool water from the single bucket of lukewarm water. Let's pour the water back into our two pitchers, then heat one on a stove while we cool the other in a refrigerator. That will certainly do it, but isn't this cheating? We get back the original temperature distribution, but only because we've introduced energy from outside sources in a selective manner. And, there is no way to introduce this energy without altering some external physical system that originally had nothing to do with our bucket of water. (Natural gas was burned in the stove, and coal was burned at a power plant to provide the electricity for the refrigerator.) What it comes down to is this: There is no way to restore order to a mixed-up system without altering some other system(s). In the context of this example, we can state the second law of thermodynamics as follows:

> The natural tendency of any isolated system is to progress from order toward disorder. Order can be created from disorder only at the expense of increasing the disorder of some other (external) system.

In this formulation, "order" refers to a configuration where one part of the system has a different temperature, pressure, color, smell, and so forth, from another part of the system. "Disorder" refers to the state where a system is so mixed up that no part is clearly distinguishable from any other part.

We get a better understanding of the water pitcher example if we think about the water molecules. Even a single drop of water contains an amazing 6×10^{21} molecules, all obviously very tiny. These molecules are in constant motion, and they travel in all directions with equal probability, at an average speed of several times the speed of sound in air. Because there are so many of them, the individual molecules travel only microscopic distances before colliding with one another, and each time they collide they exchange velocities, just like billard balls on a pool table. Experimentally, it is quite impossible to follow what a single molecule does over a period of time. Statistically, however, we find that the *average* molecular speed is directly related to the fluid's temperature, and that the average collision force is related to the pressure. In particular, we find that hot fluids have higher average molecular speeds than cold fluids.

The diagram in Figure 9.1 shows the water pitcher experiment from this molecular perspective. The cool water originally contains both fast-moving (hot) and slow-moving (cold) molecules, but the average molecular speed gives the water its observed macroscopic temperature of 10 °C. Meanwhile, the warm water also contains both fast- and slow-moving molecules, but the

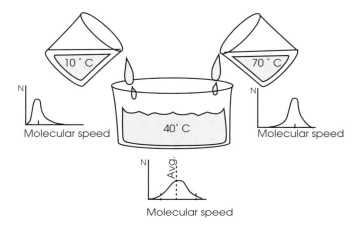

Figure 9.1. Mixing slow-moving molecules with fast-moving molecules. The equilibrium temperature reflects the final average molecular speed.

average speed is higher, and therefore the temperature is higher. Pour together the two pitchers of water, and their molecules (which are moving in random directions at extremely high speeds) quickly get mixed together. Indeed, the warm water and cool water are both still in the mixture, but their molecules have become hopelessly and irretrievably intermingled. The only observable properties of the mixture are those that relate to the statistical averages of *all* of the molecules in the mixture.

There remains a slight probability that one might measure small temperature fluctuations in the mixed water, because for some brief instant in time the slower molecules may just happen to clump together in one place while the fast ones are elsewhere. There is even a miniscule (but nonzero) probability that if you poured a liter of water out of the bucket you would happen to get just the hot molecules and leave behind the cold ones. Calculations suggest, however, that our Sun would burn out long before you had a 50–50 chance of seeing this happen. Such statistical fluctuations aside, every isolated system will evolve toward its most probable state – which is the state where everything is as randomized as it can get.

When nineteenth-century scientists learned about molecules and how they interact, it became possible to calculate the mathematical probability that these tiny constituents of matter would arrange themselves in certain patterns. The word *entropy* was coined to describe the probability that a system would evolve to a specific state as a result of its own random internal molecular motions. A low-entropy configuration was one that had a low probability of arising by chance, while a high-entropy system was one that was highly likely to develop through random molecular processes alone. In terms of entropy, the second law of thermodynamics can be stated as follows:

> The entropy of an isolated system always increases toward a maximum, at which point no further macroscopically observable physical processes are possible. To reduce the entropy of a system requires that the entropy of some other system be increased by an equal or greater amount.

Yes, this is pretty much the same as saying that systems move from states of order toward states of disorder. The fundamental difference is that entropy reflects the reason *why*. It's in the molecules.

The Universe as a whole is a closed system. (It is, by definition, all there is.) Accordingly, if the second law of thermodynamics is valid, the Universe will someday evolve to a point where all matter is uniformly distributed, all temperatures are the same, and no further energy transfers can take place, because everywhere the entropy is as great as it can be. At this point, it will be quite impossible for any life form to exist, because life forms are transient pockets of low entropy, sustained by entropy increases taking place elsewhere. When there are no suns left to burn themselves out, and when the matter of the Universe has redistributed itself into a single great sparse cloud of relatively uniform temperature and pressure, the Universe will be a boring place indeed. Some scientists believe that at that point, time itself will end.[3]

This pessimistic implication notwithstanding, the wonderful thing about the second law of thermodynamics is that in its mathematical form it permitted scientists to predict the measureable outcomes of a wide variety of thermodynamic processes, with a great degree of confidence that such predictions accurately reflected what the molecules themselves were doing. It was a tremendous shortcut. Our instruments measure only average behaviors of molecules, so why not compute with just the averages? In fact, this approach worked very effectively in the chemistry laboratory, in engine and heating system design, and in the control of a wide variety of industrial processes. It failed primarily where we already knew it was going to fail: in situations where the number of molecules was too small for the statistical approach to be meaningful. Unfortunately, this approach also failed to accurately describe large-scale natural phenomena like hurricanes, earthquakes, or epidemics. For many years the assumption was that we simply didn't have sufficient data collection technology and computational power to accurately model large complex events. There may, however, be a much more serious glitch here: It may turn out that it is not butterflies that drive the butterfly effect, but rather the statistical fluctuations in molecular distributions.

Within the last few decades, computer scientists have begun to apply the second law of thermodynamics to information systems. A low-entropy system (i.e., one with an improbable configuration) requires relatively little information to describe it. High-entropy systems, on the other hand, require a large amount of information to describe them precisely. As a system evolves from order toward disorder, the amount of information needed to describe it increases.

A simplified example will illustrate this. Suppose we have 20 molecules in a long narrow tube. Initially there are 10 molecules of substance A at the left and 10 of substance C at the right, separated by a partition. We then remove the partition and let the second law of thermodynamics take over. What information is needed to describe the configuration of the evolving system?

> Initial configuration: A A A A A A A A A A C C C C C C C C C C
> Description: *A*'s at left, *C*'s at right.

> Later configuration: A A A A A A C A C A C A C A C C C C C C
> Description: 6 *A*'s at left, 6 *C*'s at right, middle 8 alternate beginning with *C*.

> Still later: A C A A C C A C A A C A C C A C A C C
> Description: No description is possible other than to repeat the exact series.

In the third case, an approximate description might read "the *A*'s and *C*'s are randomly distributed," but even this wouldn't quite be true, because there are actually more *A*'s to the left and more *C*'s to the right of the center. Disordered systems require a large amount of information to describe them precisely, and statistical descriptions cannot possibly tell the whole story. Moreover, although it's relatively easy to see how we might create the second pattern by manipulating the first, it's increasingly difficult to hypothesize how the third pattern may have evolved from the second. In fact, there are thousands of ways this could have been done, and there are likewise millions of ways we could arrive at the third pattern while bypassing the second. The implication is this: Even if we gather great quantities of information about a complex system at one point in time, we cannot unambiguously work backward to determine the earlier states of that system. If we have a bucket of lukewarm water at 40 °C, even the most detailed information about the individual molecular speeds will not tell us how the water got to be at that temperature. If we have a hurricane, no amount of measuring after it develops will lead us to any absolute truths regarding its precursors. The mathematics of Newtonian mechanics are reversible in time, but the statistics and probabilities of the second law of thermodynamics are not.

And so, it is quite possible that the statistical fluctuations hidden within the second law of thermodynamics are in fact the butterflies in the butterfly effect. If the atmosphere were an isolated system, it would probably behave itself very well. The atmosphere is not, however, isolated from the Sun and Earth, and little pockets of low entropy are constantly being generated by infusions of solar energy. There are innumerable ways for clusters of atmospheric molecules to increase their collective entropy. If an increase in molecular entropy follows one path rather than another, it may be as if a butterfly had flitted to the left of a tree rather than to the right. Yet, the challenge of observing nature at this level of detail is far, far beyond the capability of any data collection system we can presently imagine, and it is probably well beyond the theoretical limits of what it is even possible for humans to know and computers to tally. It would appear that there needs to be a new way of looking at complex systems like the atmosphere, Earth's crust, or even the interaction of microbes with human populations.

Chaos

Toss a tennis ball into a stream above a waterfall, and try to predict the ball's velocity as it passes various points downstream. We can all picture this physical event, and many of us have done this very thing from time to time: pitching an object into a turbulent creek and curiously watching what happens. Using a conventional mathematical approach, however, the problem of predicting the motion turns out to be quite unsolvable. The turbulence below a waterfall (or in rapids) is characterized by a fluid velocity that varies wildly from point to point and from instant to instant. Not only is the fluid unruly, but it interacts with the motion of a ball that may be alternately spinning (about different axes), bobbing up and down, moving sideways, or momentarily reversing direction upstream. To complicate matters further, the buoyancy and fluid drag on the ball depend on the amount of foam in the water, and specifically on how the individual air bubbles of various sizes are distributed over the ball's submerged surface from instant to instant. Perhaps, we might argue, some incredibly patient applied mathematician might at least in principle succeed in laboring for a few years to program a computer to predict the behavior of our tennis ball in this simple example. But even this supposition would be wrong; the problem is in fact inherently unsolvable in terms that are physically meaningful.

There are analytical mathematical reasons for this, but I'll skip them for now. To see that the problem is indeed unsolvable, we can ask Mother Nature directly: that is, we can do some experiments, document what happens, then examine if there was any way we could possibly have anticipated the outcome by using traditional algebra, calculus, and/or electronic computation. Imagine that we toss some tennis balls into a river above a waterfall, then use ultrasonic sensors to monitor their changing velocities. For simplicity, let's assume that the balls don't sink permanently, nor do any get hung up on a bank. The graph in Figure 9.2 shows how the velocity measurements might vary with the downstream position for a series of identical balls. All the balls start with essentially the same velocity above the falls, but they gain slightly different velocities as they go over the falls, they land in somewhat different places, and then some immediately float free while others are caught for a long time in the turbulence. What a particular ball does is unpredictable, because so many of the environmental variations that profoundly affect its future motion are too small to be observed. What we get from our experimental graph, regardless of the precision of our measurements, is *chaos*.

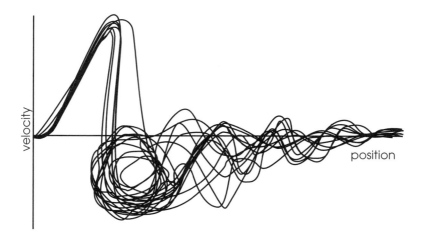

Figure 9.2. Velocities of tennis balls at points downstream from the breast of a waterfall. No two balls reproduce the same motion.

In fact, in chaos lies the great thrill that attracts enthusiasts to whitewater canoeing and rafting: It is impossible to predict what will happen from one instant to the next when you're running whitewater. Both the stream and your own motion are chaotic, and the interactions are highly complex. Yet the butterfly effect comes to the rescue of those canoeists who pay attention, for a minor correction at a critical moment can spell the difference between an exciting ride, a boring ride, or a disastrous ride.

But does our graph of Figure 9.2 really yield nothing of value? Certainly it has at least one curious feature: There are two combinations of coordinates (x, v) that all of the balls pass near, if not through. One point is in the turbulence below the falls, where some of the balls get trapped for awhile, approximately repeating a reversing upstream and downstream motion. The second point is farther downstream, where the turbulence has subsided and all of the balls again travel at about the same speed. These two points are examples of what have come to be known as *strange attractors*. From a mathematical point of view they are strange insofar as they defy analytical prediction, but from a physical point of view strange attractors usually have a quite meaningful interpretation. Chaotic systems never wander aimlessly through the universe; they always spend most of their time hovering around one of a finite number of dynamic forms.

River rapids are one example of a chaotic system. So are clouds, both when they billow slowly and when they quickly twist themselves into the frenzy of a tornado. Changing shorelines, a flag fluttering in a breeze, forest fires, avalanches, the progress of an epidemic, and movements of the earth's crustal plates are also chaotic. The time scale does not seem to be important to the description of a chaotic system, nor does the geometrical scale. Chaos can be slow or fast, its dimensions microscopic or astronomical.

Paradoxically, chaos does seem to exhibit a kind of order. This is not the order of Newtonian determinism, nor even the order of the statistical determinism of the second law of thermodynamics. Rather, it is the order of strange attractors and (perhaps at even a more fundamental level) the property of *self-similarity over scale changes*.

Go to some rapids and frame them in the viewfinder of a camera, so that there are no human figures or recognizable artificial objects in view. Now zoom in and out on different parts of the rapids and take a bunch of pictures. When you look at the finished prints, you'll find that you have no way of judging how big or small the rapids are in each picture, or how close or far the photographer stood from them. Rapids are self-similar over scale changes: Big ones look pretty much the same as small ones. Although I personally thought I had accepted this basic idea a decade ago, it was really driven home just recently as I stood near the New River Gorge Bridge in West Virginia, viewing a series of rapids in the gorge through a pair of binoculars. From my vantage point these rapids appeared to be fairly benign, whitewater that wouldn't seriously threaten an open canoe. Then a six-passenger raft came into view from around the bend, took what seemed an incredible length of time to enter the rapids, and, to my astonishment, completely disappeared from view several times as it rode the waves. Only in reference to these rafters was I able to accurately judge the impressive size of these rapids. Self-similarity over scale changes had fooled me completely.

The same is true for tornadoes, avalanches, earthquakes, and a wide variety of other natural but chaotic events: Little ones look pretty much the same as big ones. Look at an aerial photograph of a coastline or a cloud as viewed from an aircraft. Without some other recognizable reference feature in view, there is simply no way to gauge the scale. A generic coastline looks about the same whether it is photographed from a height of a meter or a thousand meters. A little cloud close-up is geometrically indistinguishable from a big cloud farther away. This suggests a fundamental shortcoming in the Newtonian deterministic approach to computational prediction: After all, if a big chaotic system is just a scaled-up version of a smaller system, why should one or the other be more difficult to describe mathematically?

Yet, self-similarity over scale also arises in some of Mother Nature's non-chaotic systems. The delight in the Japanese art of bonsai is in being tricked into a false illusion of scale: The trees look as though they're full sized, yet we can see that they're really quite small and quite alive. We aren't truly confused by the illusion, just entranced. Look at a fern: Each leaflet is an image of an entire leaf, and each leaflet is in turn made up of smaller minileaflets that have the same shape. Look at a snowflake, and, as you magnify it (assuming you can keep it from melting), you find a continuous repetition of a sixfold symmetry down to microscopic levels. Look at a photograph of a sea snail or conch, and there is no way to tell how big the creature was unless the photographer included something to provide a reference scale. These examples are not chaotic systems in the sense of rapids or tornadoes, yet they share the property of self-similarity over scale.

But ferns and snails are relatively primitive life forms, and self-similarity over scale does not normally apply to more sophisticated (i.e., lower entropy) organisms. If you see a cow or a hawk, you can pretty accurately judge how far away it is, because you know that these animals fall within a certain absolute size range. Although self-similarity over scale may be a necessary condition for chaos, then, it is hardly a sufficient condition. Chaos seems to arise only in those systems that exhibit this kind of self-similarity *and* whose dynamics are extremely sensitive to initial conditions. Strange attractors seem to be a consequence of these preconditions; they become apparent only after we observe a chaotic system doing its chaotic things for some period of time.

What really allows us to identify a chaotic system, however, is this: *Statistical predictions become meaningless.* Statistics for the average position and velocity of a drop of water in a thundercloud have no meaning, because such averages themselves change unpredictably in time, and they change differently for different droplets. Chaos is not a random process, because randomness implies that some set of outcomes occurs with fixed probability. The roll of a die is random, because each of the six faces comes up roughly one-sixth of the time. If you roll a die large number of times and average the outcomes, you will get something fairly close to 3.5. Do this again, and again the average comes close to 3.5. With chaos, however, this does not happen. Toss a tennis ball into our earlier turbulent stream and measure the time it takes to travel between two points. Do this a large number of times, and compute the average. Then repeat the process and again calculate the average time. The second average tends to be significantly different from the first. Repeating the experiment again and again doesn't help. The averages never do settle down. A few may be close to one another, then the next may be wildly divergent, for no obvious reason.

But why can't we calculate (or measure) meaningful averages of the dynamic variables of chaotic systems? Averages are a simple enough concept, and they have served us well in our use of the second law of thermodynamics. In fact, if a system has just one strange attractor, the statistics of the second law do tend to work pretty well. But increase the number of strange attractors to two, three, ten, or twenty, and averages become increasingly meaningless. To see why this is so, let's consider the chaotic system graphed in Figure 9.3. This hypothetical system has three strange attractors, which is to say that the behavior of the system may hover near one set of dynamical coordinates, then suddenly flip to one of two drastically different states where it hovers near another set of coordinates. Because the "flips" are extremely sensitive to initial conditions, they cannot be anticipated through physical measurement. As long as the system hovers near one particular strange attractor, we can expect that its behavior can be meaningfully described through averages. Unfortunately, the system will flip to the vicinity of another strange attractor without warning, and when this happens, the previously calculated average behavior is no longer descriptive. So why don't we just calculate three averages, one for each strange attractor? Because we have no way of knowing when a flip is about to occur, or how often. The best

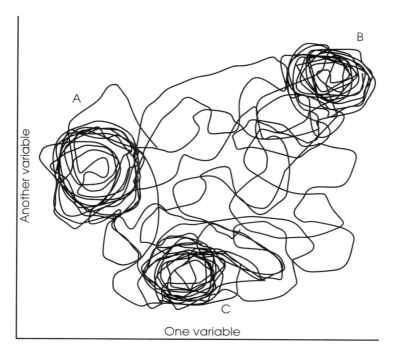

Figure 9.3. A hypothetical chaotic system with three strange attractors.

we usually can do in practice (at least at present) is to take data over some interval of time, then compute an overall average, totally ignorant of whether the system may have flipped to another strange attractor. Even, however, if we did succeed in calculating three separate averages, this information would still be of dubious value, given that we have no way of predicting *which* strange attractor the system may choose to hover about at some future instant in time.

This does not mean that people don't do statistical computations with chaotic systems. They often do, and this sometimes has seriously misleading consequences. A case in point is the insurance industry's reliance on highly formalized actuarial statistics. One can easily calculate (for instance) the average number of hurricanes that struck Florida in each of the last ten decades. One can tally the number of hurricane-related insurance claims in recent decades and relate this to the total population being insured today. On this basis, an insurance company can assess its exposure to risk from future hurricane-related claims and set aside sufficient reserve funds to meet this risk with a considerable safety margin. This would be a sound analysis if the earth's atmosphere respected the law of averages. In fact, it doesn't. In the aftermath of Hurricane Andrew in 1992, at least six Florida insurance companies were forced into bankruptcy, and many others incurred heavy losses that exceeded their financial reserves. Tropical storms, as well as patterns of tropical storms over time, are chaotic. Statistical determinism just doesn't apply to these events.

Unfortunately, modern science is still not ready to offer practical alternatives to conventional statistical analyses. Chaos theory and nonlinear dynamics are at today's scientific frontier, and current research still seems to be generating more new questions than answers. Perhaps the most compelling unanswered question is this: Is there any strategy we humans can devise to discover, *a priori*, the strange attractors of a chaotic system? If we could someday accomplish this, many futuristic social benefits might follow: weather control, earthquake control, and perhaps even epidemic control. For when a chaotic system is about to flip from the vicinity of one strange attractor to another, that is precisely when it is most sensitive to the tiny disturbances that will drive it one way or another. Why worry about our inability to predict the time and place of landfall of a hurricane if we gain the practical ability to snuff out or divert the disturbance before it ever develops into a hurricane?

Although the concept of disaster control today sounds hopelessly far-fetched, in fact there is nothing in chaos theory that renders it impossible. One way to stop or deflect a hurricane might be to produce an underwater explosion that rolls cool water to the ocean surface at a critical instant; one futuristic way to prevent earthquake casualties might be to evacuate everyone first, then trigger a series of minor earthquakes to relieve the tectonic strain. But before control must come understanding. And right now, we still don't know how to detect when a complex system is about to flip from one strange attractor to another. It is only when such a flip is about to occur that we humans may have a chance of using small amounts of energy to significantly alter the course of a large-scale chaotic phenomenon.

Transitions to Chaos

In the recorded history of mankind, the natural rise and fall of the tides has never caused a natural disaster. Even in the Bay of Fundy, where the tides roll in quickly and submerge shorelines to a depth of 12 meters (40 ft, corresponding in height to a significant tsunami), villages are never washed away, and seldom is any human caught unaware. The tides are predictable, their flow and ebb driven by an astronomical clockwork. Their behavior is the antithesis of chaos.

Natural disasters, on the other hand, are by nature unpredictable; they kill and destroy because they catch people by surprise. We grasp at the little straws Mother Nature supplies: Shoreline dwellers evacuate when the barometer plummets; ship captains steam into deeper waters at news of an incoming tsunami; Californians learn to recognize the P-wave that precedes the most devastating earthquake waves by several seconds. Yet we are also concerned about statistical prediction, and here we grasp at other straws. We pay additional money to build our houses to the prevailing earthquake or hurricane or snow-load standards, and we vaccinate our children against the most common childhood diseases. We do these things because we believe that statistically they reduce our risk (and in fact they probably do).

Yet when a natural disaster strikes, there is always a credibility gap between the scientific predictions, even the statistical ones, and the *a posteriori* reality. Natural disasters are invariably chaotic, and although over a human lifetime they sometimes appear to be statistically determinate, over longer terms even the statistics are wildly inconsistent. To speak of the "one hundred-year flood"[4] or the "fifty-year wind," regardless of the mounds of data provided to support these concepts, is to entertain a delusion. The disaster statistics of the current century are quite different from the corresponding statistics of centuries past, and future disaster statistics are bound to be different from those we tally at present. We cannot realistically predict, for instance, even the *average* number of hurricanes that can be expected annually over the following decade.

Yet we know that every physical event is grounded in the collective behavior of large numbers of molecules. Although molecular motions are random, they are definitely not chaotic. How, then, does Mother Nature escalate phenomena that are fundamentally well behaved at the molecular level into the macroscopic statistical disorder of chaos?

This is a very deep question, and the scientific answer remains incomplete.[5] If, however, we make enough careful observations, we begin to notice that chaos does not arise all at once in its full-blown complexity; rather, it evolves (sometimes slowly, sometimes rapidly) from systems that are fundamentally orderly at the onset. A simple experiment provides a bit of an insight. Turn on a sink faucet just a little, so the water barely drips out. Put an inverted pan below, so you can hear the drips. The drip frequency will be fairly uniform: "Drip . . . drip . . . drip . . ." Now increase the flow, and at first the drips come more rapidly, but still at a uniform rate. A bit more flow, however, and the dripping flips into a new pattern of two distinct frequencies: "Drip-drip . . . drip-drip . . . drip-drip . . ." Increase the flow further, and you'll get another frequency doubling: "Drip . . drip-drip . . drip . . . drip . . drip-drip . . drip . . . drip . . drip-drip . . drip . . ." By this point, the drip pattern is probably extremely sensitive to any tiny additional increase in the flow, and you may or may not be able to distinguish other patterns as the rapid dripping degenerates into chaos. Controlled experiments reveal, however, that the drip frequency goes through a whole series of bifurcations, or splits, before the regular behavior of the individual water drops is totally lost in the chaos of a turbulent stream of water.

Yet even with an ordinary faucet, you'll probably find it possible to adjust the flow to a point where the dripping water flips between two states, one where it merges into a stream, the other where it breaks into discrete drops. At this point, the tiniest touch on the faucet handle will push the system toward one of these states or the other-orderly dripping or continuous flow.

The onset of chaos usually seems to have this precursor: a series of frequency bifurcations as increasing amounts of energy are pumped into the system. At low (subcritical) energies, most systems are orderly, well behaved, and predictable. At higher (but still subcritical) energies, a system flips into more complex modes of behavior, which are still fairly well behaved. As the energy is increased further, however, it takes less and less additional energy to flip the system to the next mode of complexity. Ultimately, there is a critical amount of energy that pushes the system into chaos.

This is quite apparent in laboratory studies and computer simulations, but does it also happen in nature? Apparently so. Volcanoes, earthquakes, and even epidemics all seem to have critical points in their evolution where only the tiniest alteration in the energy balance can flip the system into one mode of behavior versus another. If we could understand and anticipate such critical points in complex phenomena, we might someday have much more success in predicting and even mitigating natural disasters.

Global Climate

Climate is not weather; it is rather the set of norms or averages about which the weather fluctuates. The climate of Rhode Island, for instance, is described in part by an average January temperature of 28.2 °F, an average July temperature of 72.5 °F, an average 124 days of precipitation per year amounting to 45.32 inches of rain and 37.1 inches of snow, and so on. No single year ever conforms very precisely to such a set of regional norms, and, as we've seen, the averages themselves drift around from decade to decade and century to century. By *global* climate, we mean the set of norms for the planet as a whole: average sea temperatures, mean air temperatures, annual cycles of expansion and contraction of the polar icecaps, the geographical distribution of rainfall, and so on.

Chaos theory throws a wrench into this conventional definition of climate. For if the behavior of the atmosphere is fundamentally chaotic, then there are no physically meaningful averages or norms. We know, for instance, that ice ages have come and gone at seemingly irregular intervals, and we find dinosaur bones in deserts where no large reptile could possibly survive today. The graph in Figure 9.4 shows what we presently know about variations in the average global temperature over the past 100,000 years; although temperature is only one of several variables that define climate, it's clear from this data that Earth's climate has not been particularly stable until relatively recently.[6] In fact, around 120,000 years ago, at a time when Earth's average temperature was very close to what it is today, there was one violent century when sea levels throughout the world rose 6 meters, then plunged more than 15 meters.[7] If a similar event were to occur today, it would mean enormous hardship for hundreds of millions of people. Moreover, climate seems to exhibit self-similarity over scale; that is, if we look at the temperature and rainfall data for a series of Januarys, this data alone won't give us much hint of whether we're looking at a specific city, a whole state, or an

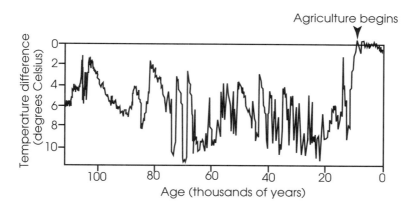

Figure 9.4. Global temperatures have fluctuated wildly until relatively recently in Earth's history. (Adapted from W. S. Broecker, Chaotic climate, *Scientific American*, Nov. 1995.)

entire country. But if climate is indeed chaotic, it ought to have strange attractors. Is there evidence that it does?

Computer simulations suggest that our global climate may indeed have at least three principal strange attractors (plus perhaps a number of minor ones). One strange attractor corresponds to the typical conditions we have grown to expect in our short human history: moderate temperatures and moderate cloud cover and rainfall. The other two strange attractors are drastically different. One is the "white Earth" attractor, where much of the planet's surface is covered with snow, there are few clouds, and overall temperatures remain quite low because the snow reflects sunlight back into space rather than absorbing it. The third possible strange attractor is the "greenhouse attractor," characterized by a thick blanket of high altitude clouds and surface temperatures high enough to evaporate the oceans. (This is what seems to have happened on the planet Venus, where today the surface temperature hovers around 500 °C, or 900 °F.) If climate is chaotic, it has the ability to flip from hovering near one of these strange attractors to bouncing around near one of the others.

Of course, human life has a vested interest in having our planetary climate remain near its current strange attractor. In fact, the chances are great that it will do so, at least over the short term. For the long term, however, all bets are off; it is quite possible that someday conditions will conspire to flip the climate into the "white Earth" mode, where living conditions will be extreme, or worse, into the "greenhouse Earth" mode, where human life will be impossible. What would trigger such a flip? Chlorofluorocarbons from refrigerants released into the atmosphere are one possibility (although there now seems to be an international effort to bring these particular emissions under control). A major meteor impact or a large volcanic explosion is another possibility. But a flip in global climate could also be induced by something we never noticed or thought of. Chaotic systems can exhibit dramatic responses to very tiny stimuli, particularly when they hover somewhere in the netherland between two strange attractors. It is certainly not prudent for us humans to alter our global environment in a manner that drives it away from its current strange attractor, for we presently haven't the foggiest notion of how far we can go before the dynamics of our climate are in danger of flipping catastrophically to another strange attractor.

It would be very helpful, of course, to have a catalogue of all of our atmosphere's strange attractors, for then we could redefine our global climate in terms of the closest strange attractor. For now, we can only rely on computer simulations, which amount to no more than a sophisticated form of speculation. The greenhouse Earth mode may be a real strange attractor, or it may be no more than an artifact of the assumptions of our computations. The mathematics and number crunching are not the reality. Only Mother Nature herself has the power to define what can be.

The Dilemma of Irreproducibility

Throughout this book, I've spoken at some length on the history of scientific thinking, and I've mentioned a few of the paradigm shifts that were needed to get science moving again when it had stalled. To Pythagoras goes credit for linking mathematics with observations of natural phenomena, Galileo we can credit with the concept of empirical investigation, and Newton gave us universal clockwork determinism. Throughout these centuries, it was reproducibility that distinguished science from nonscience: If one investigator's findings could not be replicated by another observer at another time (and perhaps place), then the initial findings were properly relegated to the trash heap.

Then, in the late nineteenth century, scientists found that many experiments and observations on atomic systems simply refused to yield to attempts at precise empirical replication (that is, replication within the limits of measurement uncertainty). This shook the foundations of science, and only after many decades did a new paradigm emerge, that of statistical determinism. Now it was okay if individual events deviated from previous observations, provided that one could reproduce the statistical measures that described the class of similar events. The paradigm of statistical determinism was quickly embraced by the more complex sciences (e.g., biology, psychology, and sociology), and science marched forward once again. Under the methodologies of statistical inquiry, new specialties like geophysics, meteorology, volcanology, and epidemiology blossomed. A paradigm shift that had arisen out of problems in understanding nature's smallest forms was now routinely applied to large-scale systems of mind-boggling complexity.

Using the paradigm of Newtonian determinism alone, it would be quite impossible to develop a science of weather, volcanoes, epidemics, or demographics. One simply cannot replicate such phenomena to the limits of measurement precision. Two volcanoes will always exhibit significantly different dynamics, as will two hurricanes, two epidemics, or the growth of two populations. The potential variables are too many, the accessible information too meager. Yet with statistics and a large enough sample size, it seemed that science could still provide insights of real value. And sure enough, we learned that subduction volcanoes generate different risks than rift volcanoes, we learned what geographical regions are most likely to be struck by hurricanes, and we learned what strategies are most effective in containing epidemics. We learned these things statistically, not analytically. And in doing this, at least some of us learned to live with our ignorance of how to predict the specific times, places, and extents of future natural disasters.

But as the twentieth century unfolded, it became increasing apparent that many complex systems simply don't yield to the paradigm of statistical determinism. Today, the methods of statistical inquiry are generating fewer and fewer really new insights into complex phenomena, and science is on the verge of another impasse, similar to the one of the 1890s. Over and over, we find that our statistical descriptions are not as transferable as we had hoped they would be; new patterns of earthquakes don't duplicate old patterns, and new diseases don't spread like previous ones, even on an average. We respond by fragmenting our sciences into pockets of increasing specificity, and in doing so we learn more and more about less and less. Except for the astrophysical theorists and the evolutionary biologists, the Newtonian concept of universality is pretty much dead in modern science.

I am not suggesting that we aren't occasionally making discoveries that in turn drive wonderful technological advances. We certainly are. But we also did this in the 1890s, a period when technological ingenuity was bursting at the seams while the physical scientists were entering a state of total confusion about the fundamental nature of matter. Technology always lags behind science, sometimes by a few years, sometimes by decades. The new wonder drugs that keep appearing on today's market are the product of scientific thinking that goes back dozens of years, and most modern electronics is based on science no younger than two or three decades. Although indeed highly specific scientific inquiry often yields the most rapid social benefits, over the longer term it takes more universal scientific insights to drive the entire process.

At present, unfortunately, we are doing precious little new science that might be drawn upon to devise novel ways of avoiding or mitigating natural disasters. Technologically, we continue to use the same time-tested methods: building seawalls, imposing construction codes, vaccinating vulnerable populations, launching new weather satellites, and so on, and our focus is on trying to do these things better rather than on developing new and different strategies. In this we have little choice, since there has been no significantly new science to guide us. But as time goes on, the challenge increases; year by year there are more pockets of high population density that are increasingly dependent on society's infrastructures and increasingly at risk from potential disasters. As we do the same old things to safeguard our populations, we may be losing ground. The devastation of Hurricane Andrew in 1992 or the Northridge earthquake of 1994 would have amounted to but a small fraction of their respective $30 billion price tags had these natural events struck twenty years earlier, when the affected regions had smaller populations.

As scientists, we are today being driven to the conclusion that we simply don't have a fruitful way of studying complex irreproducible phenomena. We are at the same kind of impasse that the physical scientists of the 1890s experienced, and which led them to the paradigm of statistical determinism and the methodologies of statistical empiricism. It now seems to be time for a new kind of science, not to replace the old science in arenas where it works, but to transcend the old science. The pockets of reality where statistical methodologies are currently fruitful will remain, just as Newtonian determinism survived the scientific revolution of the early twentieth century and continues to be used successfully to guide our space probes today. Our view then will be that Newtonian clockwork determinism is a special case of statistical determinism that applies to events where the probabilities hover near 100%, and that statistical determinism is a special case of the "new science" where probabilities are not 100% but at least are consistent over time. The "new science" will be the most general, dealing with natural events where the probability of a specific outcome is neither 100% nor consistent over time.

The main theme of this "new science," then, must be that it describes those natural phenomena that are intrinsically irreproducible, and yet does so in a way that enlightens us meaningfully about events that have not yet occurred. This is a tall order, and one that the scientific specialist is unlikely to fill, for the new science will need to transcend and merge most of our current (and artificially designated) scientific disciplines and subdisciplines. We need to seek a meaningful theory of everything complex. Is such a theory possible? I don't know. Some scientists express grave doubts, while others struggle valiantly to make progress toward this very goal.

The mathematical methods of this new science may be something quite different from what today's scientists are accustomed to. Algebra and differential equations don't seem to work well with complex systems (although they serve Newtonian determinism quite well), and conventional statistics fail when applied to chaotic phenomena. If there is to be a breakthrough, it will likely require a revolutionary new way of thinking, analyzing, and maybe even observing. Chaos theory and nonlinear dynamics offered great promise when they first emerged in the 1970s and 1980s, but, although their approaches have had great success in the computer and the laboratory (resulting in the publication of thousands of scientific journal articles), so far they have failed miserably to tell us much of real value about the outside physical world. Although the studies do generate compelling analogies between computer simulations and a wide variety of complex natural phenomena, we are left crying for answers to our more practical problems: What can this analysis tell us about next year's hurricane season, or the best allocation of equipment to fight forest fires among our national parks, or the best way to regulate casualty insurance companies? On practical questions like these most of the studies are silent, leaving our engineers and public policy planners in a total fog about what chaos theory might be capable of telling them.

It may turn out that we've already reached the natural limit of our human ability to anticipate and plan for the irreproducible events that become natural disasters. Maybe, in fact, all we can possibly do in the future is more of what we've done in the past, just with a bit more diligence. But we don't know that this is the case, and it's certainly much too soon to give up. The efforts need to continue. At this point we have no idea of where the next breakthrough might come, what obscure piece of research might provide a missing link, or what minor missed opportunity will fail to prevent a major environmental flip to an unanticipated strange attractor.

And so, let me make this appeal: The next time you hear a congressperson or senator ridiculing the money spent for research in Antarctica, or objecting to environmentalists' concerns about the burning of tropical forests, or complaining about the United Nation's birth control initiatives in the third world, or criticizing an "irrelevant" space probe to another planet, think very carefully before you jump on that person's science-bashing bandwagon. There is no way anyone can say what knowledge will be irrelevant in the long run or what seemingly obscure discovery may contribute to a major paradigm shift that will significantly alter humankind's future relationship with the natural environment.

Human civilization on Planet Earth is by no stretch of the imagination immune from potential disasters of unimaginable proportions. We are pawns in a grand scheme we have scarcely begun to understand, and we humans are far from an essential element in the future course of the Universe. We owe it to our children to negotiate a mutually respectful relationship with Mother Nature, and the only way we can do that is to listen, listen, listen, paying attention not just when she howls, but also when she whispers, ever so gently, in a language we still don't fully understand.

Notes

1 As for the first law of thermodynamics, historically it came later and identified heat as a form of energy that must be included in an accounting of the energy balance in any physical phenomenon. It is to Carnot's credit that he was able to formulate the second law without a full understanding of the nature of heat and thermal energy.

2 Rudolph Diesel had such confidence in the second law of thermodynamics that his first prototype diesel engine (1892) was two stories tall, its design completely based on second law computations. The device promptly exploded and sent Diesel to the hospital, where he modified the design during his recovery.

3 One great puzzle has always been the question of how entropy is related to time. On a molecular level, time is reversible, which is to say that you wouldn't be able to tell the difference between a videotape of a molecular interaction played forward or played backward. On a macroscopic level, however, the direction of time is obvious: If a mound of shaving cream is going *into* the can, we know that the tape is being played in reverse, because (in scientific language) entropy does not spontaneously decrease. Although a further discussion here would be a digression from the theme of this book, I strongly recommend to the interested reader Stephen Hawking's *A Brief History of Time* (New York: Bantam, 1990).

4 A "100-year flood" is one that has an alleged probability of 0.01 (1%) of occurring in a given place in a given year; statisticians themselves are careful not to say that such a flood occurs at regular intervals of 100 years. My point is that there is no way of establishing this as a meaningful number, even if one could collect data for 1,000 years and average it, for the average 100-year flood in one millennium is significantly different from the average 100-year flood in the preceding and following millennia.

5 For a discussion of some of the current philosophical issues, see John Horgan. From complexity to perplexity, *Scientific American*, June 1995, 104–9. For an earlier but more comprehensive overview of the entire subject of complexity and chaos, I recommend James Gleick, *Chaos: Making a new science* (New York: Viking, 1987).

6 W. S. Broecker, Chaotic climate, *Scientific American*, Nov. 1995, 62–8.

7 C. Stock, High tidings, *Scientific American*, Aug. 1995, 21–2.

APPENDIX A
Notable Tsunamis

Date	Location	Max. Ht. (m)	Deaths	Notes
1626 BC	Aegean Sea	u	u[a]	Following volcanic explosion at Thera.
479 BC	Greece	u	1,000s	Persian army attacking Potidea was innundated.
365, July 21	Mediterranean	u	u	Tsunamis generated by an earthquake struck entire Mediterranean coastline.
869, July 13	Japan	u	~1,000	Following local undersea earthquake.
1509, Sept. 14	Turkey	u	u	Sea rose over walls of Galata and Constantinople following earthquake.
1562, Oct. 28	Southern Chile	u	u	About 1,450 km of coastline affected.
1570, Feb. 8	Chile	u	u	Great damage reported.

1611, Dec. 2	Japan	25	3,000	Local earthquake.
1640, July 3	Japan	u	700	Following eruption of Komagatake.
1692, June 7	Port Royal, Jamaica	u	1,000s	Ships carried over city; city destroyed; capital moved to Kingston.
1703	Japan	u	100,000	Distant source, apparently in eastern Pacific.
1792	Japan	~100	15,000	Landslide; 500 million cubic meters of rock and earth fell 510 m into sea.
1746, Oct. 28	Peru	24.4	4,800	Local earthquake.
1755, Nov. 1	Lisbon, Portugal	12.2	10,000	Coastlines of Spain and northern Africa also affected.
1783, Feb 5	Scilla, Italy	u	2,473	Earthquake centered near Messina.
1811, Dec. 16	New Madrid, Mo.	u	5	Earthquake generated tsunami on Mississippi River.
1820, Dec. 29	Indonesia	21	u	Ships carried over houses.
1837, Nov. 11	Chile	5	u	Damage as far as Hilo, Hawaii.
1856, Aug. 23	Japan	u	21	Following eruption of Komagatake.
1868, Mar. 17	U.S. Virgin Islands	9.1	u	Local earthquake. Seismic activity began four months earlier and caused smaller tsunamis.
1868, Apr. 2	Hawaii	3.7	46	Local volcanic earthquake.
1868, Aug. 13	Peru	21	10,000	Caused by earthquake. Generated 4.6-meter waves in Hawaii.

(*cont.*)

Date	Location	Max. Ht. (m)	Deaths	Notes
1883, Aug. 27	Java and Sumatra	30	33,000	Volcanic explosion of Krakatau.
1883, Oct. 6	Port Graham, Alaska	7.6	u	Volcanic explosion of Mt. St. Augustin.
1896, June 15	Japan	24	26,975	Local earthquake.
1918, Oct. 11	Puerto Rico	6.1	5	Local earthquake.
1922, Nov. 11	Northern Chile	9.14	200	Local earthquake.
1927, Nov. 21	Chile	u	u	Ship with crew flung into treetops.
1929, Nov. 18	Newfoundland	15	u	Caused by earthquake at Grand Banks; heavy damage as waves swept up rivers.
1933, Mar. 3	Japan	25	2,986	Magnitude 8.9 earthquake.
1944, Dec. 7	Japan	u	998	Local earthquake.
1946, Apr. 1	Hawaii	16.5	173	Earthquake in Aleutian Islands. Height 30 m at Unimak Island.
1946, Aug. 4	Dominican Republic	5	100	Local earthquake.
1960, May 22	Southern Chile	20	2,000	Also killed 61 in Hawaii, 100 in Japan, and 20 in Philippines. Magnitude 8.3 earthquake.
1964, Mar. 27	Alaska	20	119	Magnitude 8.4 earthquake. Waves as high as 6 m in California.

1975, Nov. 29	Hawaii	4	1	Considerable damage by wave. 1 h later, Kilauea volcano erupted.
1992, Sept. 1	Western Nicaragua	10	170	Magnitude 7.0 earthquake.

Note: Deaths (fourth column) are the estimated deaths from waves alone.

[a] u = unknown

APPENDIX B

Notable Earthquakes

Date	Place affected	No. of deaths	Richter magnitude
526, May 20	Antioch, Syria	250,000	u[a]
856	Corinth, Greece	45,000	u
1057	Chihli, China	25,000	u
1290, Sept. 27	Chihli, China	100,000	u
1293, May 20	Kamakura, Japan	30,000	u
1531, Jan. 26	Lisbon, Portugal	30,000	u
1556, Jan. 24	Shaanxi, China	830,000	u
1667, Nov.	Southern Russia	80,000	u
1693, Jan. 11	Catania, Italy	60,000	u
1730, Dec. 30	Hokkaido, Japan	137,000	u
1737, Oct. 11	Calcutta, India	300,000	u
1755, June 7	Northern Persia	40,000	u
1755, Nov. 1	Lisbon, Portugal	30,000	8.75
1783, Feb. 4	Calabria, Italy	30,000	u
1797, Feb. 4	Quito, Ecuador	41,000	u
1828, Dec. 28	Echigo, Japan	30,000	u
1868, Aug. 13–15	Peru and Ecuador	40,000	u
1875, May 16	Venezuela and Colombia	16,000	u
1906, Apr. 18	San Francisco, USA	700	8.25
1908, Dec. 28	Messina, Italy	120,000	7.5
1915, Jan. 13	Avezzano, Italy	29,980	7.5
1920, Dec. 16	Gansu, China	100,000	8.6
1923, Sept. 1	Yokohama, Japan	200,000	8.3
1927, May 22	Nan-Shan, China	200,000	8.3
1932, Dec. 26	Gansu, China	70,000	7.6
1934, Jan. 15	Bihar-Nepal, India	10,700	8.4
1935, May 31	Quetta, India	50,000	7.5
1939, Jan. 24	Chile	28,000	8.3
1939, Dec. 26	Erzincan, Turkey	30,000	7.9
1946, Dec. 21	Honshu, Japan	2,000	8.4
1950, Aug. 15	Assam, India	1,530	8.7
1960, Feb. 29	Agadir, Morocco	12,000	5.8

Date	Place affected	No. of deaths	Richter magnitude
1960, May 21–30	Southern Chile	5,000	8.3
1962, Sept. 1	Northwestern Iran	12,230	7.1
1964, Mar. 27	Alaska	131	9.3
1970, May 31	Northern Peru	66,794	7.7
1976, Feb. 4	Guatemala	22,778	7.5
1976, July 28	Tangshan, China	242,000	8.2
1976, Aug. 17	Mindanao, Philippines	8,000	7.8
1978, Sept. 16	Northeastern Iran	25,000	7.7
1985, Sept. 19	Mexico City, Mexico	10,000	8.1
1988, Dec. 7	Northwestern Armenia	55,000	6.8
1990, June 21	Northwestern Iran	40,000	7.7
1995, Jan. 17	Kobe, Japan	5,250	7.2

a. u = unknown.

APPENDIX C

Notable East Coast
Tropical Storms and Hurricanes

Date	Name	Area hardest hit	Deaths in USA	Highest winds (mph)	Damage (millions)
1900, Sept. 8		Galveston, Tex.	6,000	~110	$30
1909, Sept. 21		New Orleans, La.	350	>68	5
1915, late Aug.		Texas; Louisiana	275	120	50
1915, late Sept.		Mid-Gulf Coast	275	140	13
1919, early Sept.		Gulf Coast	287	84	13
1926, mid-Sept.		Florida and Alabama	243	138	112
1928, mid-Sept.		Southern Florida	1,836	160	25
1935, early Sept.		Southern Florida	408	>150	6
1938, Sept. 21		New England	600	183	306
1944, mid-Sept.		North Carolina to New England	46	150	100

Date	Name	Region affected			
1947, mid-Sept.		Florida and mid-Gulf Coast	51	155	110
1954, late Aug.	Carol	North Carolina to New England	68	135	461
1954, early Sept.	Edna	New Jersey to New England	21	87	40
1954, early Oct.	Hazel	South Carolina to New York	95	>130	252
1955, mid-Aug.	Diane	North Carolina to New England	184	83	832
1957, late June	Audrey	Texas to Alabama	390	100	150
1960, early Sept.	Donna	Florida to New England	50	140	500
1961, early Sept.	Carla	Texas coast	46	145	408
1964, late Aug.	Cleo	Southern Florida; Virginia	3	110	129
1964, early Sept.	Dora	Northern Florida to Southern Georgia	5	125	250
1965, early Sept.	Betsy	Southern Florida; Louisiana	75	136	1,400
1967, mid-Sept.	Beulah	Southern Texas	15	100	200

(cont.)

Date	Name	Area hardest hit	Deaths in USA	Highest winds (mph)	Damage (millions)
1969, mid–Aug.	Camille	Gulf Coast to West Virginia	324	172	1,420
1970, early Aug.	Celia	Texas coast	11	130	454
1972, mid–June	Agnes	Florida to New York	118	75	2,100
1975, mid–Sept.	Eloise	Florida and Alabama	21	104	490
1979, early Sept.	David	Florida to New England	5	95	320
1979, early Sept.	Frederick	Alabama and Mississippi	5	145	2,300
1980, early Aug.	Allen	Texas coast	28	120	300
1983, mid–Aug.	Alicia	Texas coast	21	94	2,000
1985, mid–Sept.	Gloria	North Carolina; New York	8	92	1,000

| 1989, mid-Sept. | Hugo | South Carolina | 11 | 135 | 7,000 |
| 1992, late Aug. | Andrew | Southern Florida Louisiana | 58 | >155 | 32,000 |

Note: Damage estimates are based on contemporary sources, not adjusted for inflation. Storms and hurricanes before 1954 in this list had no names.

APPENDIX D

"Killer" Tornadoes

This list includes all U.S. events claiming more than fifty lives since 1925.

Date	Location	No. of Deaths	Notes
1884, Feb. 19	Ind. and the Southeast	800	About 60 tornadoes
1917, May 26–7	Ill., Ind., Ark., Ky., Tenn., Miss.	249	$5.6 M damage
1920, Apr. 20	Miss., Ala., Tenn.	220	$3.5 M damage, 6 tornadoes
1924, Apr. 29–30	Okla. and Southeast	115	$4.4 M damage, 22 tornadoes
1924, June 28	Ohio, Pa.	96	$13 M damage, 4 tornadoes
1925, Mar. 18	Mo., Ill., Ind., Ky., Tenn., Ala.	792	$17.8 M damage, 8 tornadoes
1927, Apr. 12	Rock Springs, Tex.	74	Apparently 1 tornado
1927, May 8–9	Midwest	227	$7.9 M damage, 36 tornadoes
1927, Sept. 29	St. Louis, Mo.	90	Apparently 1 tornado
1932, Mar. 21	Ala., Miss., Ga., Tenn.	321	$5.5 M damage, 27 tornadoes
1936, Apr. 5–6	Miss., Ga.	658	$21.8 M damage, 22 tornadoes
1942, March 16	Miss.	75	
1942, Apr. 27	Okla.	52	

Date	Location	Deaths	Notes
1944, June 23	Ohio, Pa., W.Va., Md.	150	$5.1 M damage, 4 tornadoes
1945, Apr. 12	Okla., Ark.	102	
1947, Apr. 9	Okla., Kans., Tex.	169	$10 M damage, 8 tornadoes
1949, Jan. 3	La., Ark.	58	
1952, Mar. 21-2	Ark., Mo., Tenn., Miss., Ala., Ky.	343	$15.3 M damage, 31 tornadoes
1953, May 11	Waco, Tex.	114	$39.5 M damage, 1 tornado
1953, June 8-9	Mich., Ohio, New England	234	$93.2 M damage, 12 tornadoes
1955, May 25	Kan., Mo., Okla., Tex.	115	$11.7 M damage, 13 tornadoes
1965, Apr. 11	Ind., Ill., Ohio., Mich., Wis.	271	$200 M damage, 47 tornadoes
1966, Mar. 3	Ala., Miss.	118	
1968, May 15	Midwest	71	$65 M damage, 7 tornadoes
1971, Feb. 21	La., Miss.	110	$17 M damage, multiple tornadoes
1974, Apr. 3-4	Ala., Ga., Tenn., Ky., Ohio	350	>$500 M damage, 44 tornadoes
1979, Apr. 10	Tex., Okla.	60	10 tornadoes
1984, Mar. 28	Carolinas	67	>$103 M damage, 30 tornadoes
1985, May 31	N.Y., Pa., Ohio, Ont.	90	43 tornadoes
1987, May 22	Tex.	29	
1989, Nov. 15	Ala.	18	
1989, Nov. 16	N.Y.	9	Single F-1 tornado
1990, June 2-3	Midwest	13	
1990, Aug. 28	Northern Ill.	25	
1991, Apr. 26	Kans., Okla.	23	55 tornadoes
1992, Nov. 21-3	South Central states	25	

Note: M = millions.

APPENDIX E

Measurement Units

Conversions between United States customary units and international units of measurement. Note: **boldface** figures are exact by international definition.

Linear measure:
1 foot = **12** inches = **0.3048** meters
1 meter = **100** centimeters = 3.280840 feet
1 kilometer = **1,000** meters = 3,280.840 feet
1 mile = **5,280** feet = **1.609344** kilometers
1 nautical mile = 1.15077 statute miles = **1,852** meters = 6076.04 feet
1 cubit (ancient Roman) = 0.444 meter

Area:
1 square inch = **6.4516** square centimeters
1 square foot = **144** square inches = **929.0304** square centimeters
1 square meter = 10.76391 square feet = 1×10^4 square centimeters
1 acre = **4,3560** square feet = 4,046.86 square meters
1 hectare = 1×10^4 square meters = 2.471 052 acres
1 square mile = **640** acres

Volume (or capacity):
1 cubic inch = **16.387064** cubic centimeters
1 liter = **1,000** cubic centimeters = 61.02374 cubic inches
1 gallon = **4** quarts = **128** fluid ounces = 3.785412 liters
1 cubic foot = 7.480519 gallons = 28.31685 liters = **1,728** cubic inches
1 cubic meter = **1,000** liters = 35.31467 cubic feet = 264.1721 gallons

Speed:
1 foot per second = **1.09728** kilometers per hour = 0.681818 miles per hour

1 meter per second = 3.280840 feet per second = 2.236936 miles per hour
1 mile per hour = **0.44704** meters per second
1 mile per hour = **1.609344** kilometers per hour
60 miles per hour = **88** feet per second

Mass:
1 kilogram = **1,000** grams = 2.204623 pounds-mass
1 pound-mass = **453.59237** grams
1 slug = 32.17405 pounds-mass = 14.5939029 kilograms
1 ton = **2,000** pounds-mass = **907.18474** kilograms
1 metric ton = **1,000** kilograms = **1** tonne

Force (including weight):
1 newton = 0.2248089 pounds
1 pound = 4.448222 newtons = **0.45359237** kilogram-force

Pressure:
1 pascal = 1 newton per square meter
1 pound per square inch = 6894.7572 pascals (Pa)
1 standard atmosphere = **101.325** kilopascals (kPa)
       = **1,013.25** millibars (mb)
       = 14.69595 lb/in^2
       = 1.0332275 kgf/cm^2
       = **760** mm Hg
       = 29.92126 in. Hg

Energy:
1 joule = **1** newton-meter = 0.737562 foot-pound
1 foot-pound = 1.355818 joules
1 kilowatt-hour = **3.6×10^6** joules
1 ton of TNT (nuclear equivalent) = 4.2×10^9 joules

Index